CONTRIBUTIONS TO MINERALIZATION

Edited by **Ali Ismail Al-Juboury**

Contributions to Mineralization
http://dx.doi.org/10.5772/intechopen.68760
Edited by Ali Ismail Al-Juboury

Contributors

Nejib Jemmali, Shaoqing Guo, Jan Rasmussen, Volker Spieth, Stanley Keith, Ali Ismail Al-Juboury, Waleed Shingaly, Elias Elias, Mohsin Ghazal, Wanming Yuan, Ke Wang, Miloš René, Yann Foucaud, Inna Filippova, Lev Filippov, Philippe Marion, Bénédicte Lechenard, M. Deniz Turan, Aneta Michalak

Published by InTech

Janeza Trdine 9, 51000 Rijeka, Croatia

Publishing Process Manager Romina Skomersic
Technical Editor SPi Global
Cover InTech Design team

First published January, 2018
Printed in Croatia
Legal deposit, Croatia: National and University Library in Zagreb

Additional hard copies can be obtained from orders@intechopen.com

Contributions to Mineralization , Edited by Ali Ismail Al-Juboury
p. cm.
Print ISBN 978-953-51-3748-1
Online ISBN 978-953-51-3749-8

For EU product safety concerns:
IN TECH d.o.o., Prolaz Marije Krucifikse Kozulić 3, 51000 Rijeka, Croatia,
info@intechopen.com or visit our website at intechopen.com.

Meet the editor

Dr. Ali Ismail Al-Juboury is professor at Geology Department of Mosul University, Iraq. He studied at Mosul University and obtained his BSc in Geology and MSc in Sedimentology in 1980 and 1983, respectively, and his PhD from Cormenius University, Slovakia, in 1992. He has published a total of 82 scientific papers in local and peer-reviewed journals in the fields of petroleum geology, sedimentology, geochemistry, and economic geology. He is a member of numerous international societies and serves in the editorial board of the Iraqi Geological Journal and International Sedimentology and Stratigraphy Journal of Oil and Gas Basins, Azerbaijan, and International Journal of Geophysics and Geochemistry (American Association of Science and Technology). He also received several awards among which are Distinguished Scholars Award from the Arab Fund for Economic and Social Development, Kuwait, in 2009 and the Science and Technology (Geology) Award from the Islamic States in 2014.

Contents

Preface

Mineralization deals with exploitation of natural resources that has a great impact on the economy and development. This book contains nine chapters that focus on genesis of selected mineral deposits hosted in carbonates or meta-sediments. These economically important mineral deposits range geographically from extreme northern Iraq to North Africa (Tunisia) and from Germany, Poland, and Portugal in Europe to China. The book also includes contributions to new trends in mineral engineering and mining industry.

Chapter 1 "Petrology, Geochemistry and Mineralogy of Greisens Associated with Tin-Tungsten Mineralization: Hub Stock Deposit at Krásno–Horní Slavkov Ore District, Czech Republic" by Miloš René introduces another example of economically important deposits of Tin-Tungsten in Europe and provides a basis for definition of the alteration zoning and estimating the magnitude of mass transfer and the integrated fluid flux involved in the formation of the stock greisen system.

Chapter 2 "Zechstein-Kupferschiefer Mineralization Reconsidered as a Product of Ultra Deep Hydrothermal, Mud-Brine Volcanism" by Stanley Keith, Volker Spieth and Jan Rasmussen presents some detailed descriptions of the world-class Kupferschiefer deposit in Poland and Germany. It is often worth presenting radical new ideas for discussion and consideration. The authors suggest exotic nature of the chemical constituents, which allows a much more flexible explanation of genetic issues, such as disparate age dates.

Chapter 3 "Lead-Zinc and Iron-Sulfide Mineralization from Northern Iraq" by Ali Al-Jaboury, Waleed Shingaly, Elias M. Elias and Mohsin M. Ghazal uses geochemical prospects through recognition, characterization, fluid inclusions and stable isotopes of the Pb-Zn-Fe sulfide ore deposits from Northern Iraq to document the mineralization stages and elucidate their paragenesis.

Chapter 4 "Geology, Textural Study, Ore Genesis and Processing of the Tabuaço Tungsten Deposit, Northern Portugal" by Yann Foucaud, Bénédicte Lechenard, Philippe Marion, Inna Filippova and Lev Filippov focuses on tungsten as an essential industrial metal with outstanding properties.

The work also presents very interesting scientific problems related to mineralogy, ore geology, and economic geology of Portugal tungsten deposit.

Chapter 5 "Lead Isotopes as Tracers of Metal Sources and Timing of the Carbonate-Hosted Pb-Zn Deposits in the Nappes Zone, Northern Tunisia" by Nejib Jemmali and Fouad Souissi examines Pb isotope data on lead-zinc deposits from the Nappes zone of Northern Tunisia to support mixing between end-member sources originating from the basement igneous rocks and the sedimentary cover rocks that are responsible for providing metals for this mineralization. This study may be useful for mineral exploration and archaeological correlation of metal artifacts.

Chapter 6 "Trace Elements in Coal Gangue" by Shaoqing Guo reviews the importance of coal gangue as one of the largest industrial residues and focuses on trace element in coal gangue that could release out and produce environmental implication in various degrees, depending on the type of trace element.

Chapter 7 "Mineralization: Evidence from Fission Track Thermochronology" by Wanming Yuan and Ke Wang discusses the application of fission track thermochronology in determining the mineralizing ages and epochs of the hydrothermal deposits.

Chapter 8 "Statistical Approach to Mineral Engineering and Optimization" by M. Deniz Turan presumed experimental data about hydrometallurgical copper extraction accompanied by the three parameters were applied to two different design models "central composite design (CCD) and Box-Behnken design (BBD)" in order to compare the results.

Chapter 9, the final chapter, "Expected Return on Capital in Mining Industry" by Aneta Michalak focuses on the mining industry. It is not only interesting for the academic field, but it is also important for investors to mining projects and mining enterprises since no economic activity of indus-trial enterprise is possible without capital.

Overall, this book provides contributions to mineralization, with case studies on selected economically exciting mineral deposits distributed in different geological ages and various geographic locations. The book also includes contributions to new aspects and techniques used in mining industry, mineral engineering, and environmental implications.

This work was achieved with great support of expert reviewers whose comments and contributions play an important and helpful role in finalizing book chapters with high quality.

We express our heartfelt gratitude to Jiří Sejkora, Khaldoun Al-Bassam, Yanfen Liao, Nejib Jemmali, Wolfgang Frisch, Majid Shahosseini, Mohsin M. Ghazal, Liu Guijian, Michal Vaněk, Sihesenkosi Nhleko, Hiyam Abass Mohammed, Adam Piestrzynski, Achim Bechtel,Dave Alderton, Gregor Borg, Stanley Keith, Sinan Akiska, Sarmad A. Ali, Farhad Ehya, Abraham JB Muwanguzi, Tomasz Niedoba, and Charles Kasanzu.

We are also indebted to InTech-Open Science in support of publishing this book.

Prof. Ali Ismail Al-Jaboury
Mosul University, Iraq

Chapter 1

Petrology, Geochemistry and Mineralogy of Greisens Associated with Tin-Tungsten Mineralisation: Hub Stock Deposit at Krásno–Horní Slavkov Ore District, Czech Republic

Miloš René

Additional information is available at the end of the chapter

http://dx.doi.org/10.5772/intechopen.71187

Abstract

The greisens evolved in the apical part of the Hub stock, formed by weakly greisenised topaz granites, are predominantly represented by Li-mica-topaz and topaz-Li-mica greisens. These greisens, relative to weakly greisenised topaz granites, are enriched in Ca, F, Fe, Li, Si, Sn and W and depleted in Al, K, Mg, Na, Ti, Y, Zr and ΣREE. Weakly greisenised topaz granites show convex tetrads in the normalised REE patterns. Compared to topaz granites, the greisens display lower ΣREE concentrations, partly higher negative Eu anomaly, high Y/Ho and low Zr/Hf ratios. Li-micas occurring in greisens are represented by zinnwaldite. Chemical composition of cassiterite is near to ideal SnO_2 (>99 wt.% SnO_2). The wolframite is represented by manganoan ferberite.

Keywords: greisen, topaz granite, petrology, geochemistry, Li-mica, cassiterite, wolframite

1. Introduction

Greisenisation is one of the most significant wall rock alterations, which occurs in granite-related tin-tungsten ore deposits [1–9]. Historically, the term "greisen" has been used firstly by miners from the Krušné Hory/Erzgebirge Mts. to describe wall rocks consisting of quartz, mica and topaz surrounding the Sn-W mineralisation [10]. This area hosts a number of Sn-W deposits (e.g., Cínovec/Zinnwald, Altenberg, Ehrenfriedersdorf, Krásno–Horní Slavkov) bound to greisenised stocks of the Variscan granitic bodies. These granites represent highly fractionated granites of the Krušné Hory/Erzgebirge batholith [11–13]. Besides the Cornwall ore fields in England, these deposits were an important European source of tin from the Bronze Age until 1991. The largest Sn-W-Li-Nb-Ta ore deposit in the Krásno–Horní Slavkov ore district is part of the Hub and Schnöd granite stocks, where Sn-W mineralisation of the greisen type was exploited from the 1200s to 1991. The total historical production of the Krásno–Horní Slavkov ore district is estimated at 52,000 tons of tin [14, 15].

This study was carried out to characterise petrological, mineralogical and geochemical features of the Hub stock greisens. The presented data provide a basis for definition of the alteration zoning and also basis for estimating the magnitude of mass transfer and the integrated fluid flux involved in the formation of the stock greisen system.

2. Geological setting

The Krásno–Horní Slavkov ore district comprises mineralised topaz granite stocks along the SE margin of the Krudum granite body in the Slavkovský les Mts. area (**Figure 1**). The Krudum granite body is part of the Western Erzgebirge pluton [16–20]. The inner structure of the granite stocks in the Krásno–Horní Slavkov ore district is remarkably stratified, comprising

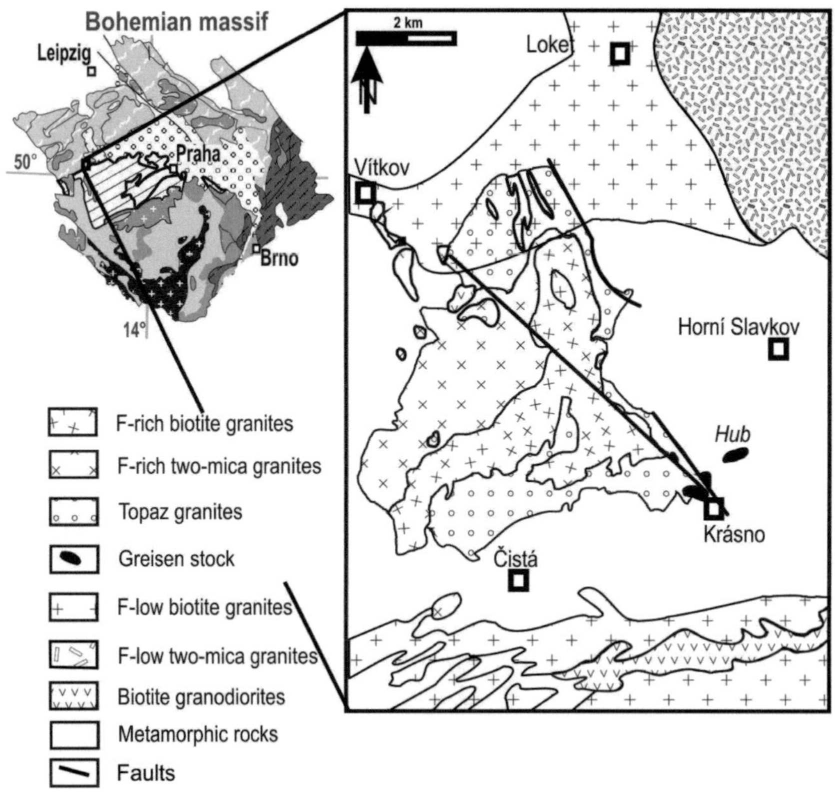

Figure 1. Geological sketch map of the Krásno–Horní Slavkov ore district.

greisens, weakly greisenised topaz alkali-feldspar granites, and layers of alkali-feldspar syenites (**Figure 2**). According to the most widely accepted genetic models, the topaz alkali-feldspar granite stocks represent the apical parts of highly fractionated granite bodies [12, 18, 19]. Greisens with Sn-W mineralisation are developed predominantly in the upper part of the Hub stock (**Figures 2** and **3**). The greisens are represented according to modal classification [2] predominantly by Li-mica-topaz and topaz-Li-mica greisens. The Li-mica and quartz greisens are less abundant. Very rare topaz greisens occur as small lenses in topaz-Li-mica greisens. Greisens occur in two structural types [19, 21]. The first and dominant one (greisens I) forms irregular bodies and lenses of various sizes. The uppermost part of the Hub stock is filled by massive greisen body, alternating deeper in the stock centres with weakly or intensively altered topaz granites. The lenticular greisen bodies are locally oriented parallel to the elongation of the stock; however, their form is usually irregular without sharp contact with weakly altered granites (**Figure 3**). The second type (greisen II) forms fracture-controlled bodies, sometimes occurring together with quartz veins (**Figure 4a**).

The occurrence of Sn-W mineralisation is controlled by the contact between granites and gneissic country rock. The highest ore concentrations occur, with some exceptions, in greisens and less frequently in the highly greisenised and argillitised topaz granites. The Sn-W mineralisation can be subdivided into (1) disseminated-type mineralisation, (2) ore domains, and (3) quartz veins. The disseminated mineralisation found in those greisens has typical content of 0.2–0.3 wt.% Sn. The ore domains are globular or even irregular bodies with tens

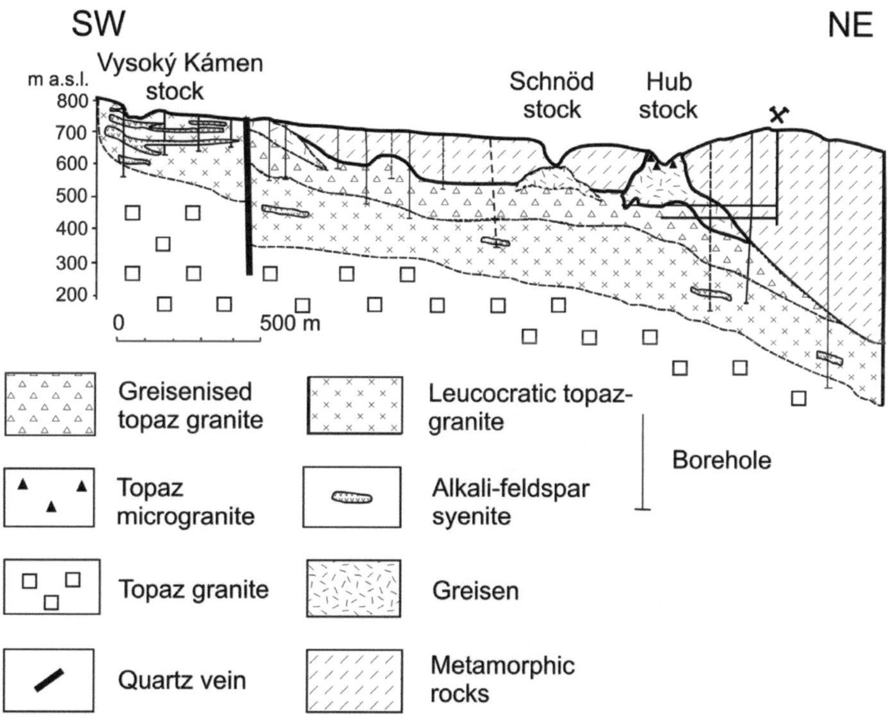

Figure 2. Schematic cross section of the south-eastern part of the Krudum granite body (after [19], modified by author).

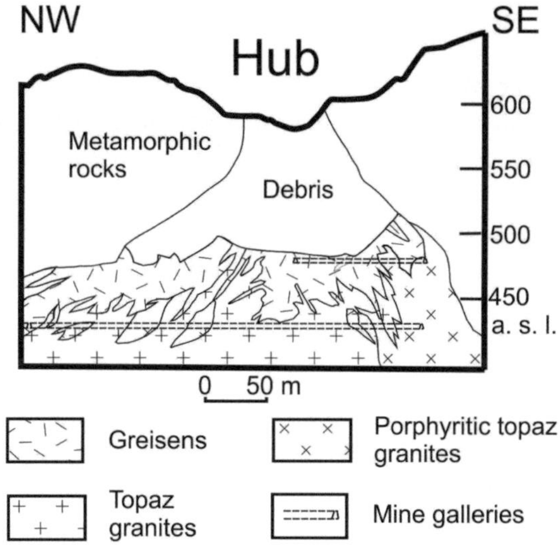

Figure 3. Schematic cross section of the hub stock (after [19], modified by author).

of centimetres in size, with a very high proportion of cassiterite (**Figure 4b**). Quartz, Li-mica, muscovite, and clay minerals (dickite, kaolinite, smectite and very rare cookeite and tosudite) are the accompanying minerals of these domains. Muscovite together with clay minerals forms a younger filling in vugs occurring in the ore domains [22, 23]. The NE–SW trending quartz veins are developed mainly in exocontact of the granite stock and less in greisen bodies, where they do not usually exceed 15 cm in thickness. In the uppermost part of the Hub stock occurs also quartz bodies with size more tens meters. Besides quartz, the quartz veins contain cassiterite and wolframite, sometimes Li-mica, apatite, and fluorite with highly variable amounts of Fe-Cu-As-Zn-Sn sulphides. The high enrichment in these sulphides especially in arsenopyrite and chalkopyrite occurs in the uppermost part of the Hub stock.

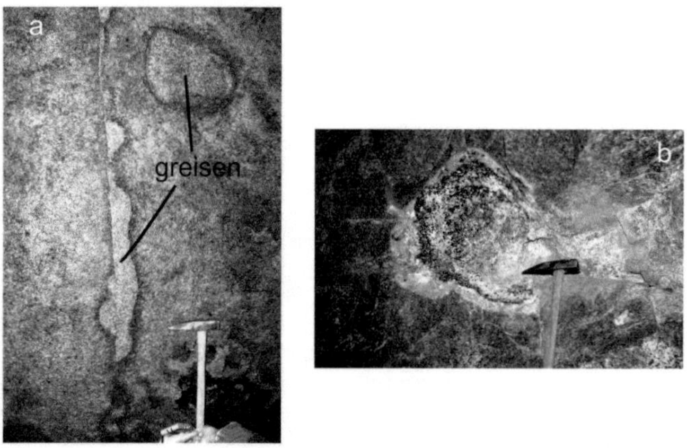

Figure 4. (a) Greisenisation II developed around the fissure in weakly greisenised topaz granite, hub stock, 4 level; (b) cassiterite-rich ore domain hosted in weakly greisenised topaz granite, hub stock, 4 level (photo of M. Košatka, 1987).

3. Samples and methods

The samples used in this study include archive material from underground mine workings and from underground boreholes performed in the area of the Hub stock. The studied greisens were sampled during the last exploration phase of this ore deposit (1987–1991). Approximately, 80 greisen and 30 topaz granite samples were used for this study. Major elements and selected trace elements (Ba, Rb, Sr, Y, Zr, Nb, Ga, and Sn) were determined by X-ray fluorescence spectrometry using the Philips PW 1410 spectrometer at chemical laboratory of the DIAMO Ltd., Stráž pod Ralskem, Czech Republic. The FeO content was measured by titration, whereas the loss on ignition (LOI) was determined gravimetrically in the same laboratory. The F content was determined using an ion-selective electrode, also at chemical laboratory of the DIAMO Ltd. For selected 10 samples, the trace elements, inclusive REE, were determined by inductively coupled plasma mass spectrometry (ICP MS) using a Perkin Elmer Sciex ELAN 6100 ICP mass spectrometer at Activation Laboratories Ltd., Ancaster, Canada. The decomposition of the rock samples for ICP-MS analysis involved lithium metaborate/tetraborate fusion. Since the analytical procedure for ICP MS involves lithium metaborate/tetraborate flux fusion, the Li concentration was analysed separately by atomic absorption spectrometry on a Varian 220 spectrometer at the Analytical laboratory of the Institute of Rock Structure and Mechanics, v.v.i. (Academy of Sciences of the Czech Republic). The density of these selected samples was determined pycnometrically at the same laboratory. Approximately 55 quantitative electron-microprobe analyses of selected minerals (Li-mica, cassiterite, and wolframite) were performed using four representative polished thin sections of greisens. Minerals were analysed in polished thin sections to obtain information about mineral zoning in examined rocks. Back-scattered electron images (BSE) were acquired to study the compositional variation of individual mineral grains. The elemental abundances of Al, Ba, Ca, Cl, F, Fe, In, K, Mg, Mn, Na, Nb, P, Rb, Sc, Si, Sn, Ta, Ti, U, W, Y, and Zr were determined using a CAMECA SX 100 electron microprobe operated in wavelength-dispersive mode at the Institute of Geological Sciences, Masaryk University in Brno. The accelerating voltage and beam currents were 15 kV and 20 or 40 nA, respectively, with a beam diameter ranging from 1 to 5 μm. Peak count-time was 20 s and background time 10 s for major elements, whereas for trace elements, 40–60 s and 20–30 s, respectively, were used. The raw data were corrected using the PAP matrix corrections [24]. The detection limits were approximately 450 ppm for Nb, 720 ppm for Ta, 600–700 ppm for U, 400–500 ppm for Y, and 600 ppm for Zr. The concentration of Li_2O in analysed micas was estimated using the following empirical equation [$Li_2O = (0.289 \times SiO_2) - 9.658$], recommended by Tischendorf et al. [25].

4. Results

4.1. Petrography

Weakly greisenised topaz alkali-feldspar granites (hight-F, high-P_2O_5 Li mica granites according to Ref. [11]) are medium-grained, equigranular rocks consisting of quartz, albite (An_{0-2}), potassium feldspar, Li-mica, and topaz. Apatite, zircon, Nb-Ta-Ti oxides, xenotime-(Y), monazite-(Ce), uraninite, and coffinite are common accessory minerals. Porphyritic, weakly greisenised topaz alkali-feldspar granites occur as relatively small lenses in the main granite body of equigranular topaz granites (**Figure 3**). Their groundmass is fine-grained with phenocrysts of potassium feldspar. Granites contain quartz, albite (An_{0-5}), potassium feldspar, Li-mica and topaz. Apatite, zircon, Nb-Ta-Ti oxides, xenotime-(Y) and monazite-(Ce) are common accessories.

Greisens I are mostly medium-grained, equigranular rocks consisting quartz, lithium mica and topaz (**Figure 5a**). Apatite, zircon, Nb-Ta-Ti-oxides, cassiterite and wolframite are common accessory minerals. Three mineralogical types of greisens were distinguished: (1) medium grained Li-mica-topaz and topaz-Li-mica greisens which prevail contain predominantly quartz (50–75 vol.%) accompanied by equant or columnar grains of topaz (9–23 vol.%) and plates of Li-mica (15–30 vol.%); (2) medium-grained quartz greisens and Li-mica greisens occur in the upper parts of the Hub stock. They are composed of 70–95 vol.% of quartz, about 5 vol.% of subhedral topaz and 5–25 vol.% of Li-mica flakes; (3) coarse-grained topaz greisens occur very rarely as small lenses enclosed in medium-grained Li-mica-topaz greisens. These greisens are also enriched in apatite. Li-mica-topaz and topaz-Li-mica greisens were partly overprinted by younger lower temperature fluids, which is evident as lithium mica together with topaz is altered to very fine-grained aggregates of muscovite (**Figure 5b**).

4.2. Whole-rock chemistry

The weakly greisenised topaz alkali-feldspar granites are a highly peraluminous rocks with an aluminium saturation index (ASI = mol.% $Al_2O_3/(CaO + Na_2O + K_2O)$ ranging from 1.1 to 1.5. In comparison with common Ca-poor, S-type granites [26], they are enriched in incompatible elements such as Rb (830–1500 ppm), Cs (38–150 ppm), Sn (19–6200 ppm), Nb (18–83 ppm), Ta (8–53 ppm) and W (4–62 ppm) but poor in Mg (0.1–0.2 wt.% MgO), Ca (0.3–1.0 wt.% CaO), Sr (12–50 ppm), Ba (21–81 ppm) and Zr (20–55 ppm). Granites are distinctly enriched in P (0.3–0.4 wt.% P_2O_5) and F (0.1–0.8 wt.% F). A highly evolved nature is reflected in the low K/Rb value (15–47).

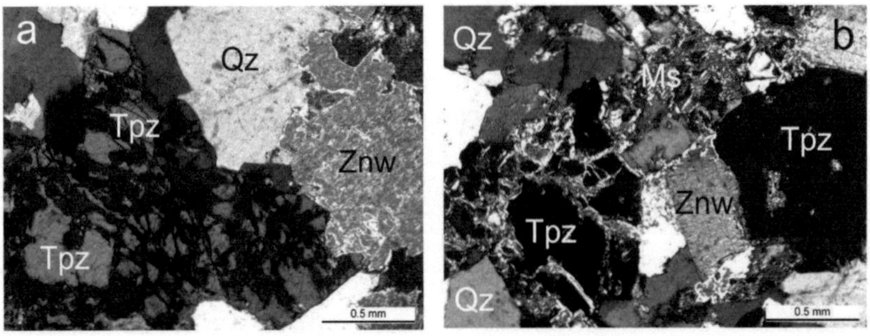

Figure 5. (a) Microphotographs of typical Li-mica-topaz greisens from the Hub stock. (a) Quartz-topaz aggregates with Li-mica flakes; (b) topaz and Li-mica altered by younger very fine aggregates of muscovite (photo of M. René, 2014).

Li-mica-topaz and topaz-Li-mica greisens are in comparison with original topaz granites enriched in Si (up to 84.4 wt.% SiO_2), Fe (0.3–4.7 wt.% $FeO_{tot.}$), F (0.1–2.8 wt.% F), Rb (201–2065 ppm), Cs (120–211 ppm), Sn (23–16,300 ppm), Nb (23–409 ppm) and W (15–160 ppm) but depleted in Na (0.2–0.6 wt.% Na_2O) and K (1.5–4 wt.% K_2O) and Zr (2–33 ppm). Li-mica greisens and especially topaz greisens are enriched in Al (up to 34.6 wt.% Al_2O_3) and depleted in Si, K, Na, Fe and Rb (**Table 1** and **Figure 6**).

Sample	999	1007	1283	1285	1000	1167	1441	1457
Rock type, wt.%	TG	TG	TG	TG	GR	GR	GR	GR
SiO_2	71.09	73.28	73.71	74.34	77.33	81.27	75.15	79.92
TiO_2	0.09	0.06	0.04	0.05	0.03	0.03	0.03	0.04
Al_2O_3	16.07	14.81	15.40	15.05	12.64	11.04	14.14	13.99
Fe_2O_3	0.10	0.04	0.01	0.06	0.10	0.58	0.25	0.07
FeO	1.08	0.91	0.77	0.81	1.30	1.90	2.73	1.95
MnO	0.08	0.08	0.04	0.05	0.12	0.14	0.17	0.13
MgO	0.23	0.17	0.05	0.05	0.05	0.05	0.05	0.10
CaO	0.53	0.60	0.40	0.38	1.57	0.34	0.77	0.36
Na_2O	3.17	3.25	3.72	3.49	0.92	0.20	0.20	0.20
K_2O	4.85	4.75	4.12	3.69	1.41	1.86	3.06	1.59
P_2O_5	0.33	0.31	0.27	0.27	0.44	0.18	0.27	0.23
H_2O^+	1.80	1.20	0.05	0.41	3.40	0.35	1.46	0.77
H_2O^-	0.00	0.00	0.00	0.32	0.00	0.50	0.42	0.26
F	0.76	0.68	0.12	0.16	2.00	1.80	0.56	0.84
O = F	0.32	0.29	0.05	0.07	0.84	0.76	0.24	0.35
Total	99.93	99.85	98.65	99.06	100.54	99.48	99.04	100.11
ppm								
Ba	55	26	23	21	22	11	28	34
Rb	1320	1030	1150	1160	732	1040	1000	971
Y	10	7	10	10	3	2	6	5
Zr	39	37	53	40	28	20	21	16
Th	5	6	9	7	2	3	4	3
La	3.54	5.06	3.46	3.77	1.83	1.33	2.35	2.13
Ce	8.09	10.32	7.91	8.70	2.12	1.47	3.62	3.26
Pr	0.95	1.21	0.91	1.03	0.23	0.17	0.43	0.39
Nd	4.01	5.43	3.47	4.22	1.08	0.84	2.16	1.95
Sm	1.38	1.38	1.07	1.39	0.31	0.25	0.65	0.59

Sample	999	1007	1283	1285	1000	1167	1441	1457
Eu	0.09	0.14	0.03	0.08	0.01	0.01	0.03	0.01
Gd	1.46	1.25	1.09	1.42	0.23	0.18	0.60	0.54
Tb	0.33	0.24	0.25	0.30	0.07	0.04	0.15	0.12
Dy	1.85	1.52	1.67	1.74	0.42	0.28	0.99	0.81
Ho	0.31	0.28	0.30	0.31	0.07	0.05	0.15	0.13
Er	1.03	0.88	0.87	0.98	0.20	0.14	0.45	0.40
Tm	0.18	0.12	0.18	0.18	0.04	0.03	0.09	0.08
Yb	1.25	0.89	1.16	1.16	0.30	0.19	0.59	0.53
Lu	0.17	0.17	0.15	0.15	0.04	0.02	0.08	0.07
La_N/Yb_N	1.91	3.82	2.02	2.20	4.12	4.73	2.69	2.72
Eu/Eu*	0.20	0.32	0.09	0.18	0.11	0.14	0.15	0.05

TG: weakly greisenised topaz granites, GR: topaz-Li-mica and Li-mica-topaz greisens, Eu/Eu* = $Eu_N/\sqrt{[(Sm_N) \times (Gd_N)]}$, normalised by chondrite using normalising value according to Ref. [28].

Table 1. Representative whole-rock chemical analyses of weakly greisenised topaz granites and greisens from the hub stock, Krásno-Horní Slavkov ore district.

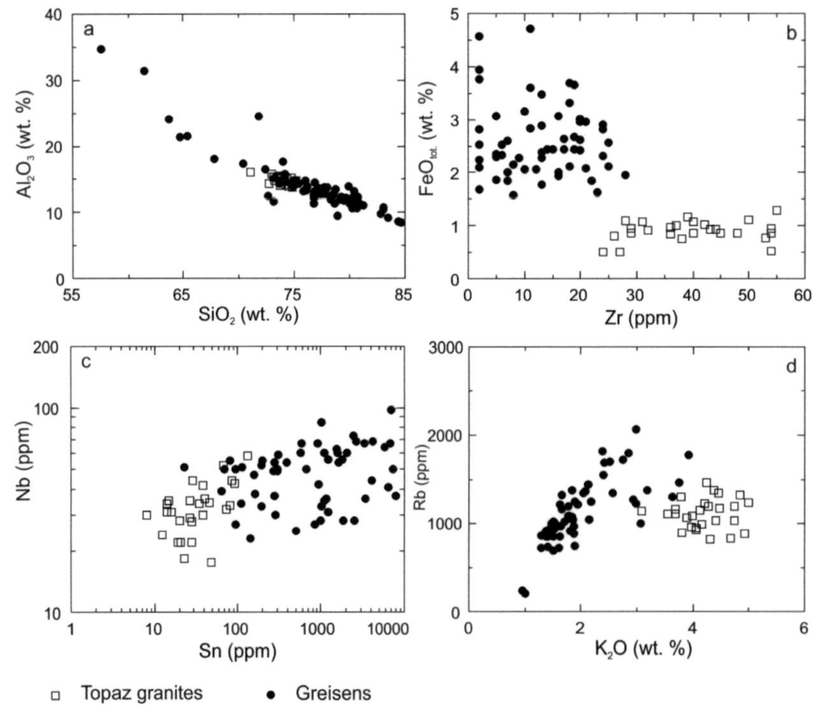

Figure 6. Whole-rock chemistry of the weakly greisenised topaz granites and different Li-mica-topaz greisens.

The normalised REE patterns indicate the decreases in the LREE and HREE in greisens compared to weakly greisenised topaz granites and steep slopes from La to Sm but almost horizontal patterns between Gd and Lu. Both rock types show prominent negative europium anomaly. An additional feature that influences the distribution of the REEs in both rock types is the lanthanide tetrad effect. Quantification of the tetrad effect sizes was carried out using method proposed by Irber [27]. Weakly greisenised topaz granites show convex tetrads in the normalised REE patterns (**Table 2** and **Figure 7**).

Rock type	Topaz granites (n = 15)	Greisens (n = 4)
Σ REE	11.17–45.97	5.00–12.33
La_N/Yb_N	1.62–4.81	2.69–4.73
La_N/Sm_N	1.61–2.41	2.27–3.71
Eu_N/Yb_N	0.02–0.51	0.05–0.13
Eu/Eu*	0.03–0.32	0.05–0.14
TE_{1-3}	1.00–1.30	0.89–1.04

Table 2. Distribution of REE in weakly greisenised topaz granites and greisens from the Hub stock.

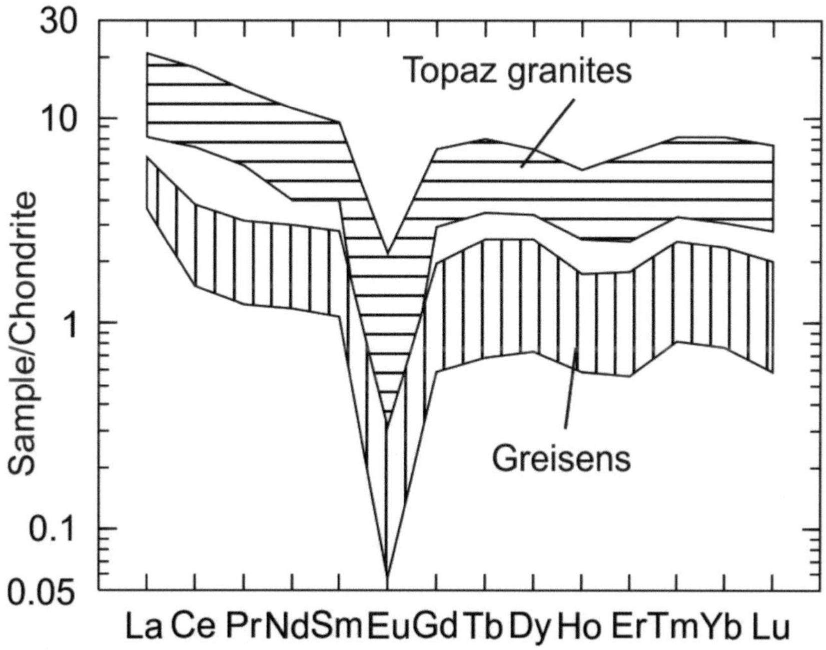

Figure 7. Chondrite normalised REE patterns of weakly greisenised topaz-albite granites and Li-mica-topaz greisens. Normalising values according to Ref. [28].

5. Mass changes during greisenisation

The origin of greisens is essentially connected to the chemical losses and gains from the alteration of the parent weakly greisenised topaz granites. For the detailed investigation of losses and gains during greisenisation, the isocon method developed by Grant [29] was applied. The scattering of the elements in the isocon plot for selected samples of Li-mica-topaz greisen suggests that the major and trace elements were mobile to variable extent (**Figure 8**). Origin of Li-mica-topaz greisen was connected with enrichment of Ca, F, Fe, Li, Mn, Si, Sn and W and loss of Al, Ba, K, Mg, Na, P, Rb Th, Y and Zr. Elements hosted by accessory zircon, monazite and xenotime (such as Zr, Th and Y) become partly depleted during greisenisation. The oxides as K_2O and Na_2O, together with Rb and Ba, were removed due to feldspar dissolution.

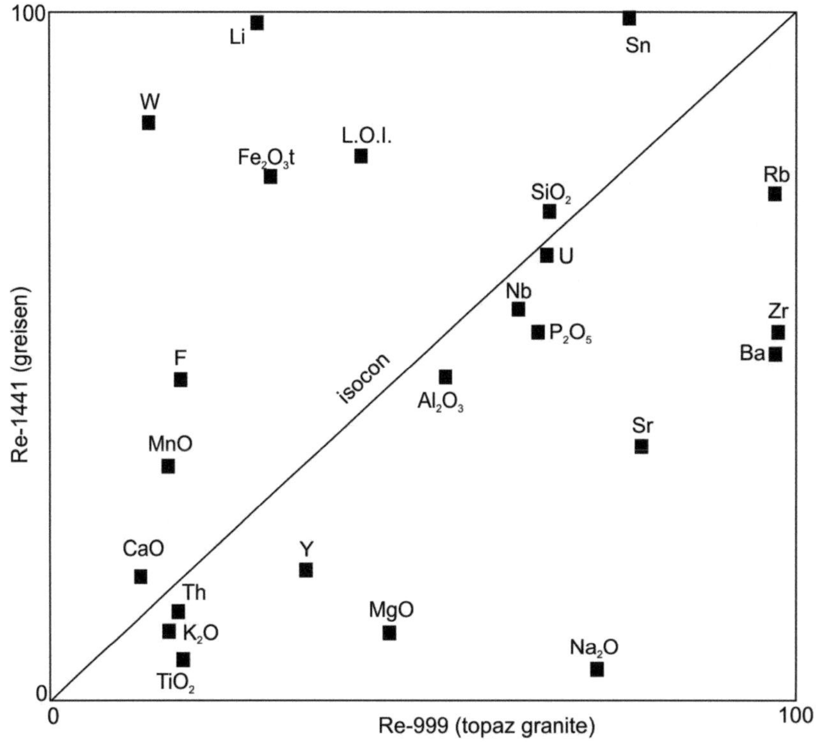

Figure 8. Normalised isocon plot of the Li-mica-topaz greisen (1441) vs. partly greisenised topaz granite (999).

6. Mineral chemistry

6.1. Lithian mica

Lithian mica in greisens according classification of Tischendorf et al. [25] is represented by Fe-rich polylithionite (**Figure 9** and **Table 3**). Grains of zinnwaldite are chemically homogeneous, without

http://dx.doi.org/10.5772/intechopen.71187

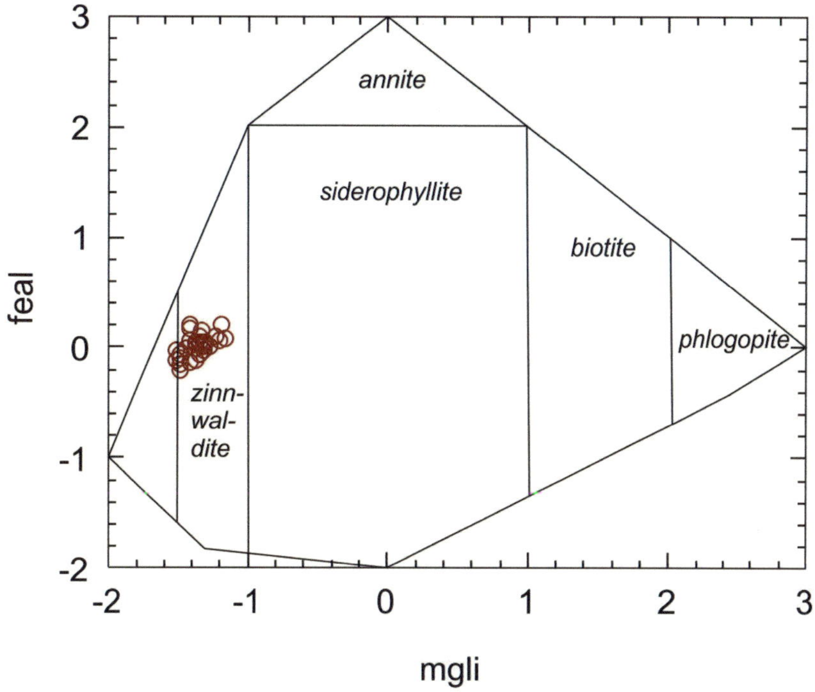

Figure 9. Subdivision of trioctahedral K-mica varieties in the mgli-feal diagram according to Ref. [26] with analyses of Li-mica from Li-mica-topaz greisens, Hub stock.

Sample, wt.%	1000-1	1000-3	1000-4	1000-7	1167-14	1167-16	1167-18	1167-20
SiO_2	42.05	42.22	42.18	41.37	41.41	42.29	41.41	41.59
TiO_2	0.47	0.18	0.07	0.27	0.30	0.05	0.26	0.45
Al_2O_3	22.06	22.56	22.28	22.38	22.45	22.29	22.50	21.73
FeO	14.64	14.38	14.74	15.14	15.08	14.98	15.15	15.81
MnO	0.63	0.60	0.66	0.64	0.81	0.79	0.72	0.66
MgO	0.13	0.13	0.11	0.13	0.04	0.04	0.04	0.03
BaO	b.d.l.	b.d.l.	0.02	0.03	b.d.l.	b.d.l.	0.02	b.d.l.
CaO	b.d.l.	0.02	b.d.l.	0.03	b.d.l.	b.d.l.	b.d.l.	b.d.l.
Na_2O	0.16	0.27	0.22	0.23	0.27	0.22	0.21	0.23
K_2O	10.12	9.97	10.40	10.06	9.97	10.20	10.07	9.95
Rb_2O	0.74	0.66	0.59	0.65	0.71	0.62	0.81	0.84
Li_2Ocalc.	2.49	2.54	2.53	2.30	2.31	2.56	2.31	2.36
F	5.16	5.31	5.79	4.98	5.47	4.83	5.07	5.01
Cl	0.01	b.d.l.	b.d.l.	0.02	0.01	b.d.l.	0.01	b.d.l.
O = F, Cl	2.17	2.24	2.44	2.10	2.31	2.03	2.14	2.11
Total	96.49	95.86	97.15	96.13	96.52	96.84	96.44	96.55

Sample, wt.%	1000-1	1000-3	1000-4	1000-7	1167-14	1167-16	1167-18	1167-20
apfu, O = 22								
Si^{4+}	6.22	6.22	6.23	6.17	6.16	6.23	6.16	6.20
Al^{4+}	1.78	1.78	1.77	1.84	1.84	1.77	1.84	1.81
Ti^{4+}	0.05	0.02	0.01	0.03	0.03	0.01	0.03	0.05
Al^{6+}	2.07	2.14	2.10	2.09	2.10	2.10	2.10	2.01
Fe^{2+}	1.81	1.77	1.82	1.89	1.88	1.85	1.89	1.97
Mn	0.08	0.08	0.08	0.08	0.10	0.10	0.09	0.08
Mg^{2+}	0.03	0.03	0.02	0.03	0.01	0.01	0.01	0.01
Ba^{2+}	0.00	0.00	0.00	0.00	0.00	0.00	0.00	0.00
Ca^{2+}	0.00	0.00	0.00	0.01	0.00	0.00	0.00	0.00
Na^{1+}	0.05	0.08	0.06	0.07	0.08	0.06	0.06	0.07
K^{1+}	1.91	1.87	1.96	1.91	1.89	1.92	1.91	1.89
Rb^{1+}	0.07	0.06	0.06	0.06	0.07	0.06	0.08	0.08
Li^{1+}	1.48	1.51	1.50	1.38	1.38	1.52	1.38	1.41
F^{1-}	1.25	1.28	1.40	1.21	1.33	1.17	1.23	1.22
Cl^{1-}	0.01	0.00	0.00	0.01	0.01	0.00	0.01	0.00
Fe/(Fe + Mg)	0.98	0.98	0.99	0.98	1.00	1.00	1.00	1.00

b.d.l.—below detection limit.

Table 3. Representative microprobe analyses and crystalochemical formulae of lithium micas from greisens, Hub stock.

internal zoning. In many places, they contain small zircon and rarely uraninite inclusions. Estimated lithium contents in analysed lithian micas from greisens reach up to 1.55 apfu (atoms per formula unit). Fluorine contents in these mica reach up to 1.40 apfu.

6.2. Sn-W minerals

Cassiterite occurs usually as subhedral to anhedral chemically homogenous grains in the BSE (backscatter electron) mode. Its chemical composition is near ideal SnO_2 (>99 wt.% SnO_2). The only important trace elements are Ti (up to 0.28 wt.% TiO_2), Fe (up to 0.16 wt.% FeO) and W (up to 0.63 wt.% WO_3). Contents of Ta and Nb are low, up to 0.32 wt.% Ta_2O_5 and up to 0.23 wt.% Nb_2O_5, respectively (**Table 4**).

Wolframite occurs here generally as relatively rare anhedral grains without discernible chemical zoning. The wolframite is represented by manganoan ferberite with 70–76 mol.% $FeWO_4$ (**Table 5**). Analysed wolframite is partly enriched in Nb (up to 0.72 wt.% Nb_2O_5 and Ta (up to 0.18 wt.% Ta_2O_5) with Ta/(Ta + Nb) ratios ranging from 0.11 to 0.50.

Sample	1167-43	1457-3	1457-4	1457-5	1457-8	1457-11	1457-12	1457-13
Ta_2O_5	0.02	0.02	b.d.l.	b.d.l.	0.03	0.03	b.d.l.	b.d.l.
Nb_2O_5	0.08	0.09	b.d.l.	b.d.l.	0.02	b.d.l.	b.d.l.	b.d.l.
TiO_2	0.28	0.19	0.03	0.01	0.02	0.08	b.d.l.	b.d.l.
SnO_2	99.09	99.51	99.25	99.00	99.74	100.04	99.79	100.13
WO_3	0.05	b.d.l.	0.18	0.45	0.00	0.01	0.02	0.13
FeO	0.26	0.08	0.09	0.03	0.01	0.05	0.16	0.02
MnO	0.01	0.01	b.d.l.	b.d.l.	0.01	b.d.l.	b.d.l.	0.01
Total	99.79	99.89	99.55	99.49	99.82	100.19	99.97	100.28
O = 2								
apfu								
Ta^{5+}	0.00	0.00	0.00	0.00	0.00	0.00	0.00	0.00
Nb^{5+}	0.00	0.00	0.00	0.00	0.00	0.00	0.00	0.00
Ti^{4+}	0.01	0.00	0.00	0.00	0.00	0.00	0.00	0.00
Sn^{4+}	0.99	0.99	1.00	1.00	1.00	1.00	1.00	1.00
W^{6+}	0.00	0.00	0.00	0.00	0.00	0.00	0.00	0.00
Fe^{2+}	0.01	0.00	0.00	0.00	0.00	0.00	0.00	0.00
Mn^{2+}	0.00	0.00	0.00	0.00	0.00	0.00	0.00	0.00

b.d.l.—below detection limit.

Table 4. Representative microprobe analyses and structural formulae of cassiterite from greisens of the Hub stock (wt.%).

Sample	1167-28	1167-29	1167-30	1441-36	1141-37
WO_3	75.76	76.22	76.69	75.98	75.85
Ta_2O_5	0.14	0.15	0.11	0.10	0.18
Nb_2O_5	0.67	0.72	0.19	0.04	0.23
SnO_2	0.02	0.04	b.d.l.	b.d.l.	0.02
Bi_2O_3	b.d.l.	0.04	b.d.l.	b.d.l.	0.03
Sc_2O_3	0.01	0.02	b.d.l.	0.01	0.01
As_2O_3	b.d.l.	0.01	0.01	0.01	b.d.l.
In_2O_3	b.d.l.	0.08	b.d.l.	0.08	b.d.l.
FeO	16.69	16.88	16.36	17.29	17.61
MnO	6.50	6.39	6.92	5.59	5.48
CaO	0.01	0.02	0.01	0.02	b.d.l.
Total	99.80	100.61	99.50	99.12	99.48

Sample	1167-28	1167-29	1167-30	1441-36	1141-37
O = 4					
apfu					
W^{6+}	0.99	0.99	1.00	1.00	1.00
Ta^{5+}	0.00	0.00	0.00	0.00	0.00
Nb^{5+}	0.02	0.02	0.00	0.00	0.01
Sn^{4+}	0.00	0.00	0.00	0.00	0.00
Bi^{3+}	0.00	0.00	0.00	0.00	0.00
Sc^{3+}	0.00	0.00	0.00	0.00	0.00
As^{3+}	0.00	0.00	0.00	0.00	0.00
In^{3+}	0.00	0.00	0.00	0.00	0.00
Fe^{2+}	0.70	0.71	0.69	0.74	0.75
Mn^{2+}	0.28	0.27	0.29	0.24	0.24
Ca^{2+}	0.00	0.00	0.00	0.00	0.00

b.d.l.—below detection limit.

Table 5. Representative microprobe analyses and structural formulae of wolframite (wt.%).

7. Discussion

7.1. Mineral and chemical changes during greisenisation

The greisenisation is a typical hydrothermal alteration accompanying the origin of rare-metal granites and associated Sn-W-Li-Mo mineralisation [1, 7, 30–33]. Greisens of the Hub stock are formed in the uppermost part of a weakly greisenised topaz granite cupola. In a similar position occur greisen bodies in the Altenberg, Cínovec/Zinnwald, Sadisdorf and Ehrenfriedersdorf Sn-W ore deposits, representing the most significant Sn-W deposits in the Krušné Hory/Erzgebirge ore district [34].

In the Cínovec/Zinnwald granite cupola in the Eastern Erzgebirge greisen bodies consist of two structural types: (i) flat thin greisen zones and quartz veins with wall-rock greisens, (ii) irregular greisen bodies several tens of metres in size, following the morphology of the granite cupola contact [35]. Similar greisen bodies, formed in the Nejdek-Eibenstock and Horní Blatná plutons in the Western Erzgebirge, show a pattern of wall-rock alteration, the intensity of which decreases outward. All these structural types of greisen bodies display varying mineralogical composition with predominance of quartz-, topaz-, Li-mica-topaz and Li-mica greisens in the German part of the Western Erzgebirge and with predominance of quartz-, topaz-Li-mica- and Li-mica greisens in the German part of the Eastern Erzgebirge [2]. Therefore, the mineralogical compositions of greisens (I) found in the Hub stock are similar to those of the Western Erzgebirge.

Chemical changes during greisenisation of topaz granites are strongly dependent on the prior state of the rock and the intensity of alteration. Silica in topaz greisens is lost, whereas in Li-mica-topaz and quartz greisens, SiO_2 is gained. Alumina is gained in topaz-Li-mica and in topaz greisens. Due to the decomposition of albite and alkali feldspars, there is a significant loss of Na and K. In contrast, additions of Li reflect the growth of Li-mica during greisenisation, especially in Li-mica greisens. Similar gains and losses were found in selected greisens from the German part of the Krušné Hory/Erzgebirge Mts [2].

The REE plots indicate a similar course of REE patterns in weakly greisenised topaz granite and both greisen varieties, suggesting a partial dissolution of the refractory accessory minerals, but not their complete removal. An additional feature that influences the distribution of the REE in weakly greisenised topaz granites and greisens from the Hub stock is the occurrence of the convex lanthanide tetrad effect. Older studies of the tetrad effect in fractionated granites from the Smrčiny/Fichtelgebirge and Krušné Hory/Erzgebirge Mts. proposed that processes of fluid-melt interaction caused the development of a convex tetrad effect during crystallisation of the silicate melt [27, 31]. However, more recent studies [36, 37] from the Zinnwald ore deposit in the Eastern Krušné Hory/Erzgebirge Mts. demonstrated that tetrad effect was developed prior to greisenisation in stage of sub-solidus albitisation of topaz granites. The lower ΣREE concentrations in greisens according to Ref. [37] suggested that the greisenisation resulted in remobilisation of the REEs. The depletion of the ΣREE concentrations in greisens from the Hub stock is accompanied by changes in the tetrad effect, Y/Ho, Eu/Eu* and Zr/Hf ratios (**Figure 10**). The partly increasing Y/Ho ratio in greisens could be explained by migration of fluorine-rich hydrothermal fluids [38, 39]. The more negative Eu anomaly in the greisens relative to weakly greisenised topaz granites is behaving as a divalent species during greisenisation, as predicted by Sverjensky [40] and Wood [41]. Similar increases of the negative Eu anomaly in the greisens were found in the True Hill granite greisens of southwestern New Brunswick, Canada [7]. The decreasing of Zr/Hf ratio in the greisens relative to weakly greisenised topaz granites is, according to Irber [27], affected by strong hydrothermal alteration.

7.2. Sn-W minerals

The Sn in cassiterite is usually substituted by Fe, Mn, Ta, Nb and W [42]. The enrichment of Ta and Nb in cassiterite is significant for those occurring in pegmatites, granites, and high-temperature quartz veins [43–45]. The cassiterite analyses from granites and pegmatites fall along a linear array with Nb + Ta and Fe + Mn—concentrations varying while mostly maintaining a ratio of 2:1 corresponding to the coupled substitution 3 $Sn^{4+} \Leftrightarrow 2(Nb, Ta)^{5+} + (Fe, Mn)^{2+}$. This relationship is common to cassiterite from highly fractionated granites and pegmatites worldwide and different to those from hydrothermal cassiterite-quartz-vein deposits and greisen deposits [44]. Although cassiterite is main ore mineral in greisens, data about Nb, Ta, and W concentrations are rare [46–50]. Cassiterites from greisens are usually distinctly depleted in Fe, Nb, and Ta in comparison to cassiterites from granites, pegmatites, and high-temperature quartz veins. This depletion is also significant for cassiterites from examined greisens. The depletion in Nb, Ta, and Fe could be explained by lower crystallisation temperature of greisen cassiterites [48].

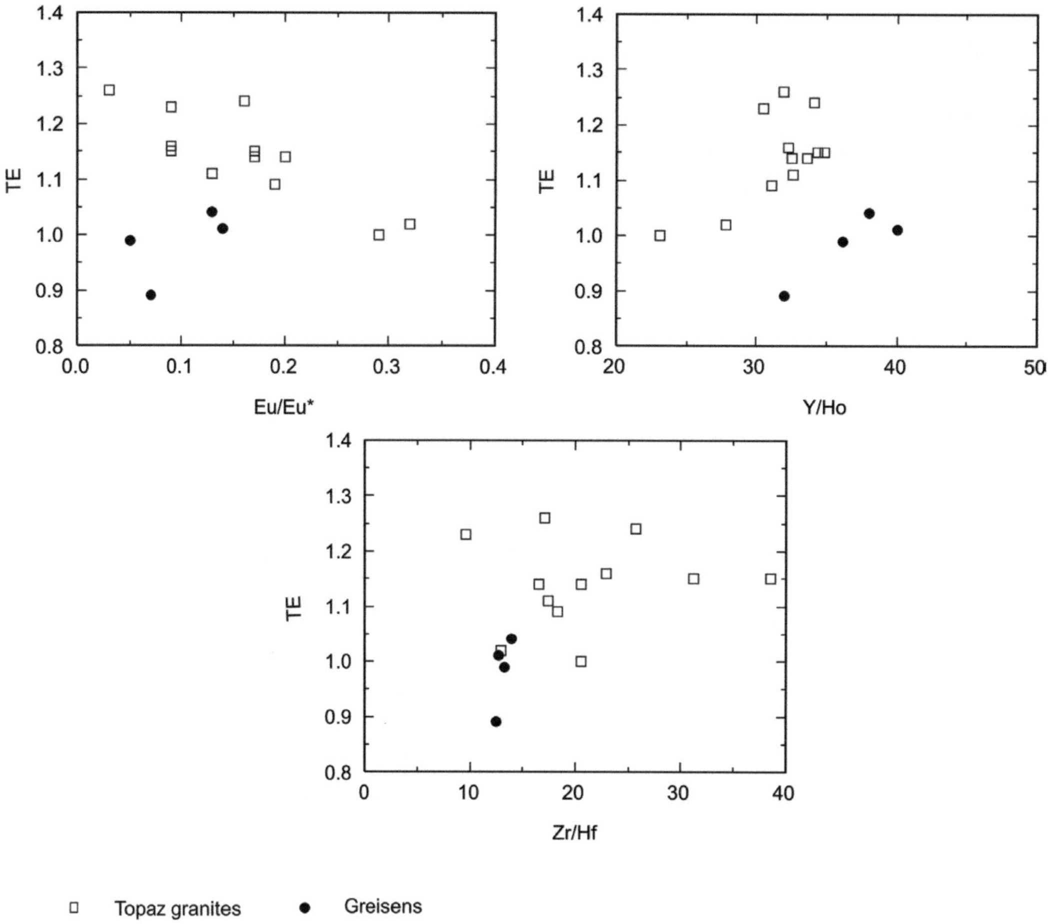

Figure 10. Evolution of tetrad effect, Y/Ho, Eu/Eu* and Zr/Hf ratios in the Li-mica-topaz greisens from the Hub stock.

Wolframite is solid solution between end members ferberite ($FeWO_4$) and hübnerite ($MnWO_4$) and shows a wide variety of its chemical composition. The reasons for compositional variability of Mn, Fe, and Nb in wolframite have been controversial for many years [51–56]. The empirical concept that the Mn/Fe ratios of the wolframite may vary according to the crystallisation temperature was disproved by the empirical study of wolframite and hydrothermal experiments [51, 54–57]. More recently, it was established that the wolframite compositions are determined by the compositions of the surrounding rocks, source materials, and by the Mn/Fe ratio of hydrothermal solutions [57–62]. The substitution of Nb and Ta in wolframites is supported by following substitution equation: $(Fe, Sc)^{3+} + (Nb, Ta)^{5+} \Leftrightarrow (Fe, Mn)^{2+} + W^{6+}$ [63]. The Nb and Ta are concentrated in wolframites from rare-metal granites and pegmatites, with predominance of Nb about Ta [64]. Wolframites from Sn-W deposits in the German part of the Krušné Hory/Erzgebirge ore district, including wolframites from the Hub stock, are ferberites with variable Mn/Fe ratios [65–67]. From quartz veins in the upper part of the Hub stock was also described the purest known (end-member) hübnerite. Hübnerite occurs in small cavities in these quartz veins [14].

8. Conclusion

The greisens of the Krásno–Horní Slavkov Ore District evolved in the apical part of the weakly greisenised topaz granite Hub stock and are represented by Li-mica-topaz, topaz-Li-mica, quartz, Li-mica and topaz greisens. These greisens occur in two structural types. The first and dominant one (greisen I) forms massive greisen bodies in the uppermost part of the Hub stock. The second type (greisen II) forms fracture-controlled bodies, sometimes occurring together with quartz veins. The disseminated Sn-W mineralisation occurring in greisens has typical content of 0.2–0.3 wt.% Sn. The predominately Li-mica-topaz and topaz-Li mica greisens relative to weakly greisenised topaz granites are enriched in Si, F, Fe, Li, Si, Sn and W and depleted in Al, K, Mg, Na, Ti, Th, Y, Zr and ΣREE. Weakly greisenised topaz granites show convex tetrads in the normalised REE patterns. The greisens display lower ΣREE concentrations, high negative Eu anomaly, high Y/Ho, and low Zr/Hf ratios relative to the topaz granites. Li-micas occurring in greisens are represented by zinnwaldite. The chemical composition of cassiterite is near ideal SnO_2 (>99 wt.% SnO_2). The wolframite is represented by manganoan ferberite with 70–76 mol.% $FeWO_4$.

Acknowledgements

This work was carried out thanks to the support of the long-term conceptual development research organisation RVO: 67985891. The microprobe analyses were performed thanks to the financial support of the Czech Science Foundation (project No. 205/09/0540). I am grateful to R. Škoda, Š. Benedová and P. Gadas from the Institute of Geological Sciences of the Masaryk University (Brno) for technical assistance with microprobe analysis of selected minerals. I wish also to thank A. Szameitat for her constructive remarks and English corrections and book editor A. I. Al-Juboury for his very useful remarks and recommendations.

Author details

Miloš René

Address all correspondence to: rene@irsm.cas.cz

Institute of Rock Structure and Mechanics, v.v.i., Academy of Sciences of the Czech Republic, Prague, Czech Republic

References

[1] Rundquist DV, Denisenko VK, Pavlova IG. Greisen Deposits: Ontogenesis, Phylogenesis. Moscow: Nedra Press; 1971. 328 p (in Russian)

[2] Kühne R, Wasternack J, Schulze H. Postmagmatische Metasomatose im Endo-Exokontakt des jüngeren postkinematischen Granite des Erzgebirges. Geologie. 1972;**21**:494-520

[3] Štemprok M. Greisenization (a review). Geologische Rundschau. 1987;**76**:169-175. DOI: 10.1007/BF01820580

[4] Richardson JM, Bell K, Watkinson DH, Blenkinsop J. Genesis and fluid evolution of the east Kemptville greisen-hosted tin mine, southwestern Nova Scotia, Canada. In: Stein HJ, Hannah JL, editors. Ore-Bearing Granite Systems; Petrogenesis and Mineralizing Processes. Geol. Soc. Amer. Spec. Pap. Vol. 246. 1990. p. 181-203

[5] Schwarz MO, Surjono A. Greisenization and albitization at the Tikus tin–tungsten deposit, Belitung, Indonesia. Economic Geology. 1990;**85**:691-713. DOI: 10.2113/gsecongeo.85.4.691

[6] Pirajno F. Hydrothermal Mineral Deposits. 6th ed. New York: Springer Verlag; 1992. 732 p

[7] Lentz DR, Gregoire C. Petrology and mass. Balance constraints on major-, trace-, and rare-earth-element mobility in porphyry-greisen alteration associated with the epizonal True Hill granite, southwestern New Brunswick, Canada. Journal of Geochemical Exploration. 1995;**52**:303-331. DOI: 10.1016/0375-6742(94)00059-K

[8] Zharikov VA, Rusinov VI, Marakushev AA. Metasomatism and Metasomatic Rocks. Nauchnyi Mir: Moscow; 1998. 492 p (in Russian)

[9] Somarin AK, Ashley P. Hydrothermal alteration and mineralisation of the Glen Eden Mo-W-Sn- deposit: A leucogranite-related hydrothermal system, southern New England Orogen, NSW, Australia. Mineralium Deposita. 2004;**39**:282-300. DOI: 10.1007/s00126-003-0399-3

[10] Lempe JF. Beschreibung des Bergbaues aus dem Sächsischen Zinnwalde. Mag. Bergbaukunde. 1785:1

[11] Förster H-J, Tischendorf G, Trumbull RB, Gottesmann B. Late-collisional granites in the Variscan Erzgebirge, Germany. Journal of Petrology. 1999;**40**:1613-1645. DOI: 10.1093/petroj/40.11.1613

[12] Breiter K. Nearly contemporaneous evolution of the A- and S-type fractionated granites in the Krušné hory/Erzgebirge Mts., Central Europe. Lithos. 2012;**151**:105-121. DOI: 10.1016/j.lithos.2011.09.022

[13] Štemprok M, Blecha V. Variscan Sn-W-Mo metallogeny in gravity picture of the Krušné hory/Erzgebirge granite batholith (Central Europe). Ore Geology Reviews. 2015;**69**:285-300. DOI: 10.1016/oregeorev.2015.02.014

[14] Beran P, Sejkora J. The Krásno Sn-W ore district near Horní Slavkov: Mining history, geological and mineralogical characteristics. Journal of the Czech Geological Society. 2006;**51**:3-42

[15] Beran P, Jangl L, Majer J, Suček P, Wagenbreuth O. 1000 Years of Tin Mining in the Slavkovský les Mts. Sokolov: Okresní muzeum Sokolov; 1996. 195p (in Czech)

[16] Fiala F. Granitoids of the Slavkovský (Císařský) les Mountains. Sbor. geol. Věd G. 1968; **14**:93-160

[17] René M. Development of topaz-bearing granites of the Krudum massif (Karlovy Vary pluton). Acta Universitatis Carolinae Geologica. 1998;**42**:103-109

[18] Breiter K, Förster H-J, Seltmann R. Variscan silicic magmatism and related tin-tungsten mineralization in the Erzgebirge-Slavkovský les metallogenic province. Mineralium Deposita. 1999;**34**:505-521. DOI: 10.1007/s001260050217

[19] Jarchovský T. The nature and genesis of greisen stocks at Krásno, Slavkovský les area - western bohemia, Czech Republic. Journal of the Czech Geological Society. 2006;**51**:201-216

[20] Machek M, Roxerová Z, Janoušek V, Petrovský E, René M. Petrophysical and geochemical constraints on alteration processes in granites. Studia Geophysica et Geodaetica. 2013;**57**:710-740. DOI: 10.1007/s11200-013-0923-6

[21] Košatka M. Geological Evolution of the Hub Stock, Central Part on the Krásno Ore Deposit [Thesis]. Prague: Charles University; 1988 (in Czech)

[22] Melka K, Košatka M, Zoubková J. The occurrence of dioactahedral chlorite in greisen. In: Proceedings of the 7th Euroclay Conference; Dresden. 1991. p. 757-760

[23] Štastný M, René M. Argillization of topaz-bearing granites in the hub stock, Horní Slavkov-Krásno Sn-W ore district (bohemian massif, Czech Republic). Acta Geodynamics et Geomaterialia. 2014;**11**:255-267. DOI: 10.13168/AGG.2014.0008

[24] Pouchou JL, Pichoir F. "PAP" (ϕ-ϱ-Z) procedure for improved quantitative microanalysis. In: Armstrong JT, editor. Microbeam Analysis. San Francisco: San Francisco Press; 1985. p. 104-106

[25] Tischendorf G, Rieder M, Förster H-J, Gottesmann B, Guidotti CV. A new graphical presentation and subdivision of potassium micas. Mineralogical Magazine. 2004;**68**:649-667. DOI: 10.1180/0026461046840210

[26] Chappell BW, Hine R. The Cornubian batholith: An example of magmatic fractionation on a crustal scale. Resource Geology. 2006;**56**:203-244. DOI: 10.1111/j.1751-3928.2006.tb00281.x

[27] Irber W. The lanthanide tetrad effect and its correlation with K/Rb, Eu/Eu*, Sr/Eu, Y/Ho and Zr/Hf of evolving peraluminous granite suites. Geochimica et Cosmochimica Acta. 1998;**63**:489-508. DOI: 10.1016/s0016-7037(99)0027-7

[28] Boynton WV. Geochemistry of the rare earth elements: Meteorite studies. In: Henderson P, editor. Rare Earth Element Geochemistry. Vol. 1984. Amsterdam: Elsevier; 1984. p. 63-114

[29] Grant JA. The isocon diagram–a simple solution to Gresens equation for metasomatic alteration. Economic Geology. 1986;**81**:1976-1982. DOI: 10.2113/gsecongeo.81.8.1976

[30] Wright JH, Kwak TAP. Tin-bearing greisens of mount Bischoff, northwestern Tasmania, Australia. Economic Geology. 1989;**84**:551-574. DOI: 10.2113/gsecongeo.84.3.575

[31] Halter WE, Willliams-Jones AE, Kontak DJ. The role of greisenization in cassiterite precipitation at the east Kemptville tin deposit, Nova Scotia. Economic Geology. 1996;**91**:368-385. DOI: 10.2113/gsecongeo.91.2.368

[32] Dolejš D, Štemprok M. Magmatic and hydrothermal evolution of Li-F granites: Cínovec and Krásno intrusions, Krušné hory batholith, Czech Republic. Bulletin of the Czech Geological Survey. 2001;**76**:77-99

[33] Štemprok M, Dolejš D. Fluid focusing, mass transfer and origin of fracture-controlled greisens in the Western Krušné hory granite pluton, Central Europe. Zt. Geol. Wiss. 2010;**38**:207-234

[34] Schust F. On the importance of granite elevations in the localization of greisen bodies in the Erzgebirge (Mts.). In: Fedak J, editor. The Current Metallogenic Problems of Central Europe. Vol. 1976. Warsaw: Wydawnictwa Geologiczne; 1976. p. 189-206

[35] Štemprok M, Novák JK, David J. The association between granites and tin-tungsten mineralization in the eastern Krušné hory (Erzgebirge), Czech Republic. In: von Gehlen K, Klemm DD, editors. Mineral Deposits of the Erzgebirge/Krušné Hory (Germany/Czech Republic). Berlin: Gebrüder Borntraeger; 1994. p. 97-129

[36] Monecke T, Dulski P, Kempe U. Origin of convex tetrads in rare element patterns of hydrothermally altered siliceous igneous rocks from Zinnwald Sn-W deposit, Germany. Geochimica et Cosmochimica Acta. 2007;**71**:335-353. DOI: 10.1016/j.gca.2006.09.010

[37] Monecke T, Kempe U, Monecke J, Sala M, Wolf D. Tetrad effect in rare element distribution patterns: A method of quantification with application to rock and mineral samples from granite-relalted rare metal deposits. Geochimica et Cosmochimica Acta. 2002, 2002;**66**:1185-1196. DOI: 10.1016/s0016-7037(01)00849-3

[38] Bau M, Dulski P. Comparative study of yttrium and rare-earth element behaviours in fluorine-rich hydrothermal fluids. Contributions to Mineralogy and Petrology. 1995;**119**:213-223. DOI: 10.1007/BF00307282

[39] Bau M. Controls on the fractionation of isovalent trace elements, in magmatic and aqueous systems: Evidence from Y/Ho, Zr/Hf, and lanthanide tetrad effect. Contributions to Mineralogy and Petrology. 1996;**123**:323-333. DOI: 10.1007/s004100050159

[40] Sverjensky DA. Europium redox equilibria in aqueous solution. Earth and Planetary Science Letters. 1984;**67**:70-78. DOI: 10.1016/0012-821X(84)90039-6

[41] Wood SA. The aqueous geochemistry of the rare-earth elements and yttrium. Part 2. Theoretical predictions of speciation in hydrothermal solutions to 350°C at saturated water vapor pressure. Chemical Geology. 1990;**88**:99-125. DOI: 10.1016/0009-2541(90)90106-H

[42] Möller P, Dulski P, Szacki W, Malow G, Riedel E. Substitution of tin in cassiterite by tantalum, niobium, tungsten, iron and manganese. Geochimica et Cosmochimica Acta. 1988;**52**:1497-1503. DOI: 10.1016/0016-7037(88)90220-7

[43] Murciego A, Garcia Sanchez A, Dusausoy Y, Martin Pozas JM, Ruck R. Geochemistry and EPR of cassiterites from the Iberian Hercynian massif. Mineralogical Magazine. 1997; **61**:357-365

[44] Tindle AG, Breaks FW. Oxide minerals of the separation rapids rare-element granitic pegmatite group, northwestern Ontario. The Canadian Mineralogist. 1998;**36**:609-635

[45] Abdalla HM, Matsueda H, Obeid MA, Takahashi R. Chemistry of cassiterite in rare metal granitoids and associated rocks in the Eastern Desert, Egypt. Journal of Mineralogical and Petrological Sciences. 2008, 2008;**103**:318-326. DOI: 10.2465/jmps.070528a

[46] Oen IS, Korpershoek HR, Kieft C, Lustenhouwer WJ. A microprobe study of rutile, cassiterite and wolframite and sulfides in the Morro Potosí greisen, Rondonia, Brazil. neues jahrbuch für mineralogie monatshefte. 1982;**4**:175-191

[47] Binde G. Beitrag zur Mineralogie, Geochemie und Genese des Kassiterits. Freiberg. Forsch.-H. 1986;**C411**:1-60

[48] Haapala I. Magmatic and postmagmatic processes in tin-mineralized granites: Topaz-bearing leucogranite in the Eurajoki Rapakivi granite stock, Finland. Journal of Petrology. 1997;**38**:1645-1659. DOI: 10.1093/petroj/38.12.1645

[49] Sala M. Geochemische und mineralogische Untersuchungen an alterierten Gesteinen aus dem Kuppelbereich der Lagerstätte Zinnwald (Osterzgebirge) [thesis]. Freiberg: University of Freiberg; 1999

[50] Somarin AK. Ore mineralogy and mineral chemistry of the Glen Eden Mo-W-Sn greisen-breccia system, eastern Australia. Journal of Mineralogical and Petrological Sciences. 2009;**104**:339-355. DOI: 10.2465/jmps.070929

[51] Leutwein F. Die Wolframit Gruppe. Freiberger Forschungshefte. 1951;**C8**:8-19

[52] Schröcke H. Isomorphiebeziehungen in der Wolframitgruppe. Beiträge zur Mineralogie und Petrographie. 1960;**7**:166-206

[53] Schröcke H. Heterotype Mischbarkeit zwischen Wolframit- und Columbitgruppe. Beiträge zur Mineralogie und Petrographie. 1961;**8**:92-110

[54] Taylor RG, Hosking KFG. Manganese-iron ratios in wolframite, south Crofty Mine, Cornwall. Economic Geology. 1970;**65**:47-53. DOI: 10.2113/gsecongeo/65.1.47

[55] Moore F, Howie RA. On the application of the hübnerite:Ferberite ratio as geothermometer. Mineralium Deposita. 1978;**13**:391-397. DOI: 10.1007/BF00206572

[56] Willgallis A. Zur Mischkristallverhältnis von Wolframiten. Neues Jahrbuch für Mineralogie Abhandlungen. 1982;**145**:308-326

[57] Hsu LC. The stability relations of the wolframite series. American Mineralogist. 1976; **61**:944-955

[58] Barabanov VF. Geochemistry of tungsten. International Geology Review. 1971;**13**(3):332. DOI: 10.1080/00206817109475439

[59] Groves DI, Baker WE. The regional variation in composition of wolframites from Tasmania. Economic Geology. 1972;**67**:362-368. DOI: 10.2113/gsecongeo.67.3.362

[60] Amossé J. Variations in wolframite composition according to temperature, at Borralha, Portugal, and Enguyales, France. Economic Geology. 1978;**73**:1170-1175. DOI: 10.2113/gsecongeo.73.6.1170

[61] Nakashima K, Watanabe M, Soeda A. Regional and local variations in the composition of the wolframite series from SW Japan and possible factors controlling compositional variations. Mineralium Deposita. 1986;**21**:200-206. DOI: 10.1007/BF00199801

[62] Campbell A, Petersen U. Chemical zoning in wolframite from San Cristobal, Peru. Mineralium Deposita. 1988;**23**:132-137. DOI: 10.1007/BF00206662

[63] Černý P, Ercit TS. Mineralogy of niobium and tantalum: Crystal chemical relationships, paragenetic aspects and their economic implications. In: Möller P, Černý P, Saupé F, editors. Lanthanides, Tantalum and Niobium. Berlin: Springer Verlag; 1989. p. 28-79

[64] Tindle AG, Webb PC. Niobian wolframite from Glen Gairn in the eastern highlands of Scotland: A microprobe investigation. Geochimica et Cosmochimica Acta. 1989;**53**:1921-1935. DOI: 10.1016/0016-7037(89)90313-X

[65] Kempe U, Wolf D. Die Interpretation des H/F-Koeffizienten – immer noch offen. Freiberg. Forsch.-H. 1994;**C451**:53-87

[66] Hösel G, Hoth K, Jung D, Leonhardt D, Mann M, Meyer H, Tägl U. Das Zinnerz-Lagerstättengebiet Ehrenfriedersdorf/Erzgebirge. Bergbau in Sachsen. 1994;**1**:1-196

[67] Weinhold G. Die Zinnerz-Lagerstätte Altenberg (Osterzgebirge). Bergbau in Sachsen. 2002;**9**:1-73

Chapter 2

Zechstein-Kupferschiefer Mineralization Reconsidered as a Product of Ultra-Deep Hydrothermal, Mud-Brine Volcanism

Stanley B. Keith, Volker Spieth and
Jan C. Rasmussen

Additional information is available at the end of the chapter

http://dx.doi.org/10.5772/intechopen.72560

Abstract

The Kupferschiefer is a copper-, polymetallic-, hydrocarbon-bearing black shale of the lowermost Zechstein Group of Permo-Triassic age (252 Ma) in Germany and Poland. It is usually 1 m thick and underlies 600,000 km^2, extending from Great Britain to Belarus for a distance of over 1500 km. At a district scale, copper has been mined for over 800 years since its discovery circa 1200 A.D. Mineralogical, chemical, and geological analyses of the combined Zechstein-Kupferschiefer show strong chemical and paragenetic relationships between the Zechstein salines, Kupferschiefer, and Weissliegend sandstones that lead to a broader, more unified, genetically linked model related to deep-sourced, hot, hydrothermal, mud-brine volcanism. The overall Zechstein-Kupferschiefer chemical stratigraphy suggests density-/composition-driven fractionation of deep-sourced, metal-rich, alkali-rich, silica-aluminum-rich, halogen-rich, high-density brines. The ultimate brine source is interpreted to be serpentinized peridotite in the lower crust near the Moho transition to the mantle. Dehydration of the serpentinite source to talc (steatization) by mantle heat during failed, intra-continental rifting of the Pangaea supercontinent at the end of Permian time released vast amounts of element-laden, high-density brines into deep-basement fractures, depositing them into and above the Rotliegend Sandstone in the shallow Kupferschiefer Sea, which is analogous to the modern northern Caspian Sea.

Keywords: Kupferschiefer, Permian, copper, silver, hydrocarbons, hydrothermal, ultra-deep hydrothermal, UDH, Zechstein, mud volcanism, exotic

1. Introduction

The Kupferschiefer is copper-bearing black shale of lowermost Zechstein Group of Permo-Triassic age (252 Ma) in Germany and Poland. The black shale is approximately 1-m thick and underlies 600,000 km^2 of northern Europe, extending from Poland to eastern England (**Figure 1**). The Kupferschiefer is thinly laminated, bituminous, calcareous and clayey shale at the base of the Zechstein Group. The Kupferschiefer is a small part of a larger, Permo-Triassic, genetically linked geosystem that includes the underlying Rotliegend (red sandstone) and Weissliegend (white sandstone), Kupferschiefer (black copper shale), and overlying Zechstein chemical sedimentary rocks (Werra carbonates, dolomite, anhydrite, and saline rocks) (**Figure 2**) that are cross cut by the hematitic Rote Fäule. In this chapter, this broader system (**Figure 3**) is designated the Zechstein-Kupferschiefer system, following the precedent of Kucha [1].

The literature that has accumulated since Agricola is extremely robust and this chapter does not attempt to summarize the pre-existing genetic models. Instead, this chapter offers a broader perspective and hypothesis that the entire Weissliegend-Kupferschiefer-Zechstein system can be reconsidered as product of ultra-deep, high-energy, high temperature, mud volcanism. These rocks are produced by a deep-sourced, hydrothermal plume system of high-density brines, and mud slurries that transported the above materials from a steatized serpentinite reaction chamber near the base of the crust to a low-energy, low temperature, exhalative system at the crust-hydrosphere interface.

The ultra-deep hydrothermal (UDH), mud-volcanic model resolves a number of paradoxes that have arisen in the abundant Kupferschiefer literature. One main contrast in concepts is between high temperature, hydrothermal mineral assemblages, and low temperature mineral

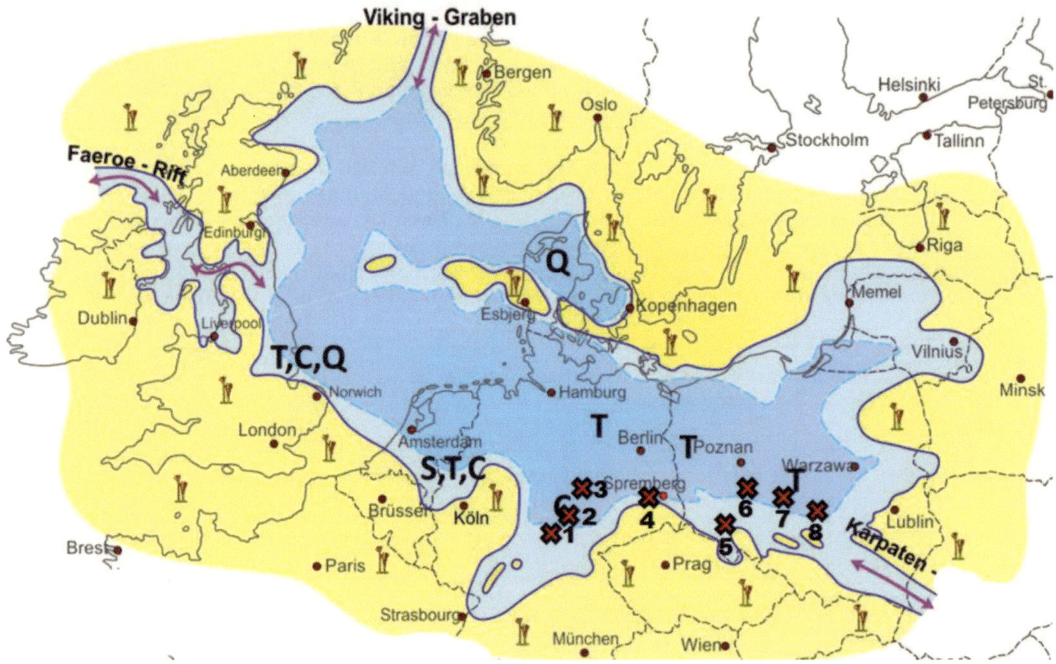

Figure 1. Map of Zechstein basin showing locations of exotic magnesium-rich mineral suites and euhedral quartz [T = talc; S = serpentine; C = clinochlore; Q = euhedral quartz]. From west to east (left to right], the locations are Yorkshire, England [TCQ], Emsland, southwest Germany [STC], Mors diapir, northern Denmark [Q], Gorbelen salt dome north central Germany [T], Königschall-Hindenburg salt mine, southern Germany [C], dolomite 'reef'. Mines: 1 = Melsungen, 2 = Sangerhausen, 3 = Mansfeld, 4 = Spremberg, 5 = Konrad, 6 = Polkowice-Sieroszowice, 7 = Rudna, 8 = Lubin.

http://dx.doi.org/10.5772/intechopen.72560

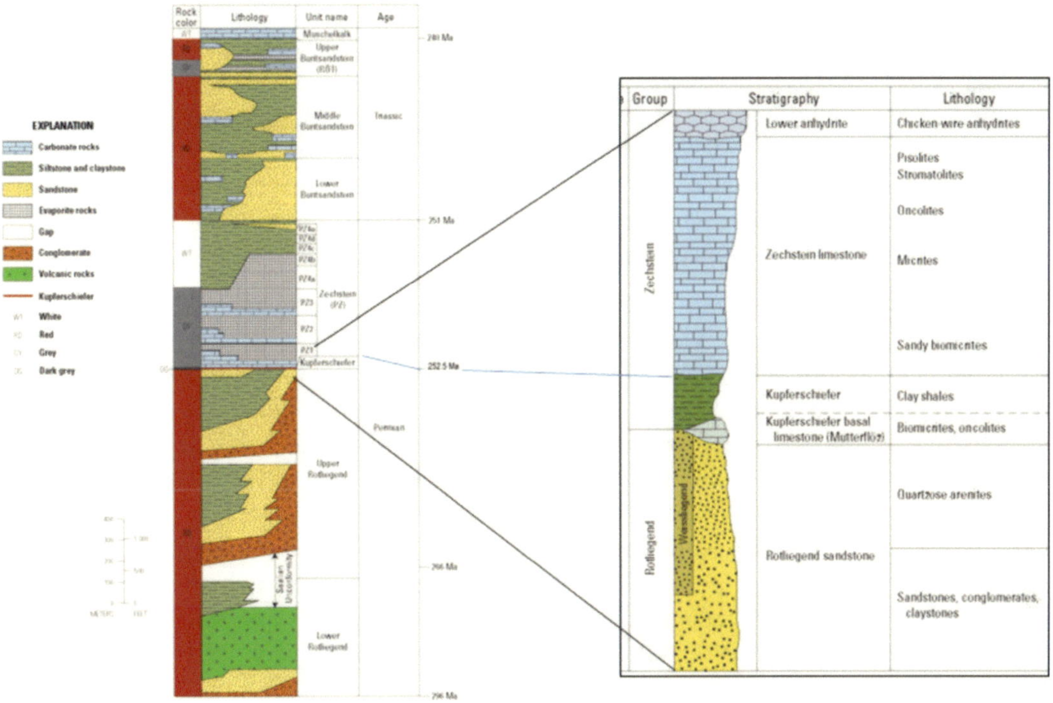

Figure 2. Stratigraphic sections (broader area on left, narrow zone on right) (modified from Zientek et al. [2]).

Figure 3. The upper Permian-lower Triassic sequence is from upper (Sandsteinschiefer) Rotliegend sandstone in the footwall, the mineralized Weissliegend white sandstone and Kupferschiefer metallized black shale, and the hanging wall of Zechstein dolostone. Location: Roerigschacht, Wettelrode/Sangerhausen mine, Mansfeld district, Germany. The Permo-Triassic boundary is placed at the Kupferschiefer-Zechstein contact. (Photo by Juergen Kopp showing the mine manager Dipl.-Ing. Erich Hartung).

assemblages indicative of surface environments. The UDH mud volcanism model merges both syngenetic and epigenetic processes. The UDH model also considers the oxidized and reduced mineralization styles to be independent, age distinct components of an ongoing, pulsed process that occurred over a period of geologic time from 265 to 235 Ma. Within this sequence, each pulse represents a thermodynamically distinct episode that is not rigorously connected to the other episodes. The connection only occurs at the scale of the entire system that is broadly related to a process of dehydration of ultramafic serpentinites in the lower crust that were episodically affected by influxes of mantle heat.

The UDH, mud-volcanic concept posits that much of the Kupferschiefer-Zechstein represents metalliferous slurry flows that were erupted from numerous mud vents. A spectacular example of where epigenesis meets syngenesis is present at Rudna where chalcocite veins transgress metallized Weissliegend and appear to erupt chalcocite slurry flows into the overlying Kupferschiefer. The slurries can contain exotic, hydrothermal components as clasts that may have originally formed deep within the mud-volcanic system at various times and temperatures. This explains the equivocal results from dating individual clasts within the Kupferschiefer. Once erupted onto the seafloor (circa 255 Ma), the deep-sourced mud slurries were able to interact with the low-energy, biogenic environment. In some cases, especially in the distal pyrite environments, microbial reduction of the sulfur could have produced light sulfur isotopes. This process, however, would have been superimposed on already very light sulfur isotopes inherited from deep sources.

The southern rim of the Kupferschiefer carries enriched polymetallic mineralization near large shear zones (**Figure 4**). Mineralization occurs as: (1) large amounts of copper sulfides in the upper part of the Weissliegend sandstones, (2) reduced, black to dark gray, kerogen-rich, metal sulfide-bearing deposits of the Kupferschiefer and overlying Zechstein carbonates, and (3) oxidized, reddish, hematitic rocks (Rote Fäule) that contain gold and platinum group elements (PGE). Mineralization consists of fine disseminations and as coarse-grained aggregates, lenses, streaks, and veinlets of sulfide minerals. Mineralization of the Rote Fäule transgresses all three units at very low angles (**Figure 5**).

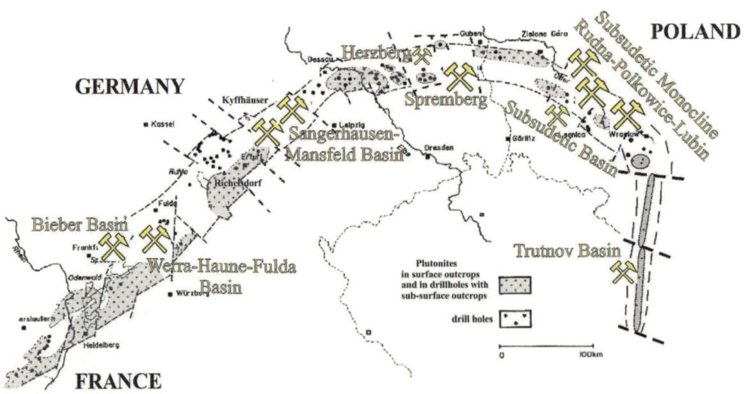

Figure 4. The copper-polymetallic-gold belt of central Europe showing exploration drill holes, deposits and mines (names of the mines are in **Figure 1**). The mineralized belt overlies the post-Variscan-crystalline zone at the southern rim of the Zechstein-Kupferschiefer Sea [62].

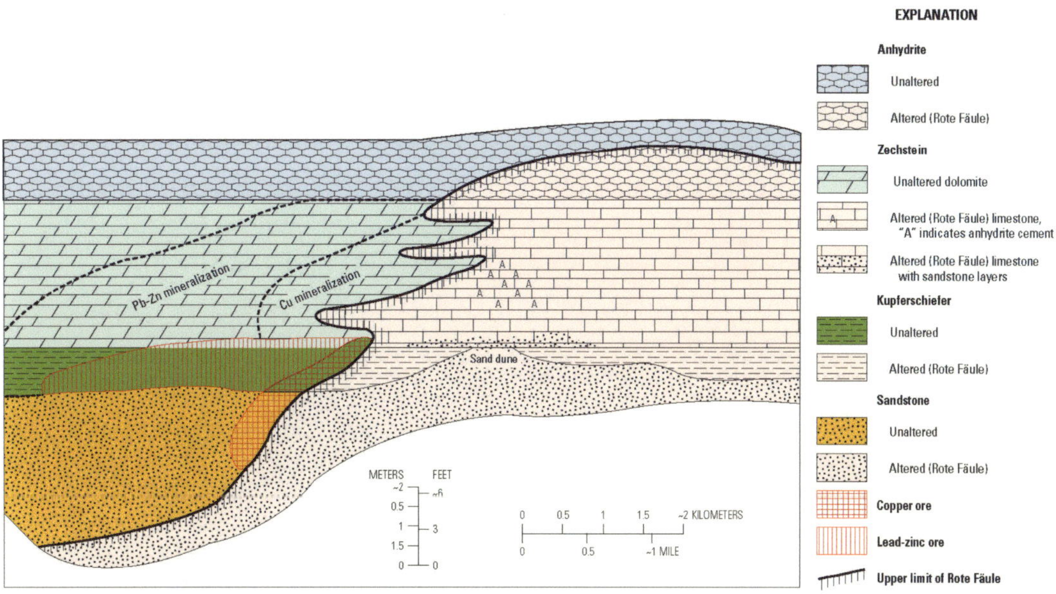

Figure 5. Schematic cross-section showing the Zechstein-Kupferschiefer metallization as part of the greater Zechstein sequence in the Lubin district (from Zientek et al. [2]).

Since about 1200 A.D., some 78 million metric tons of copper metal have either been produced or are delineated as reserves in Germany and Poland, according to an assessment of undiscovered copper resources for the Kupferschiefer (as documented as production and reserves by Zientek and others [2]). Some of the significant mines are shown in **Figure 1**. Zientek and others [2] estimate that 126 million metric tons of copper metal may remain undiscovered.

2. Formations underlying the Kupferschiefer

2.1. Rotliegend Group

The Lower Rotliegend consists of volcanic rocks and interbedded fluvial and lacustrine sedimentary strata of Early Permian age. The Upper Rotliegend consists of clastic, red bed sedimentary rocks of Middle and Late Permian time that were deposited in desert environments over broad areas of northern Europe. These rocks extend from Poland to western England and from the German-Danish border to northeast of Frankfurt, Germany [2]. The Rotliegend is approximately 300 m thick [1]. The abundant areas of oil and gas fields in the Rotliegend (**Figure 6**) (one [Groningen] of which is world class [3]), follow a similar regional orientation as the metallic mines and are partly hosted in the same units (especially pure silica sand of the Weissliegend). The economic significance of the oil and gas resource overshadows the economic significance of the metal production in the Zechstein-Kupferschiefer.

The Rotliegend underlying the black shale of the Kupferschiefer is inferred to be a pre-amble to the ultra-deep hydrothermal (UDH) process. The UDH process of the Zechstein-Kupferschiefer super system is likely to have subjected the Rotliegend rocks to alteration along feeder zones and

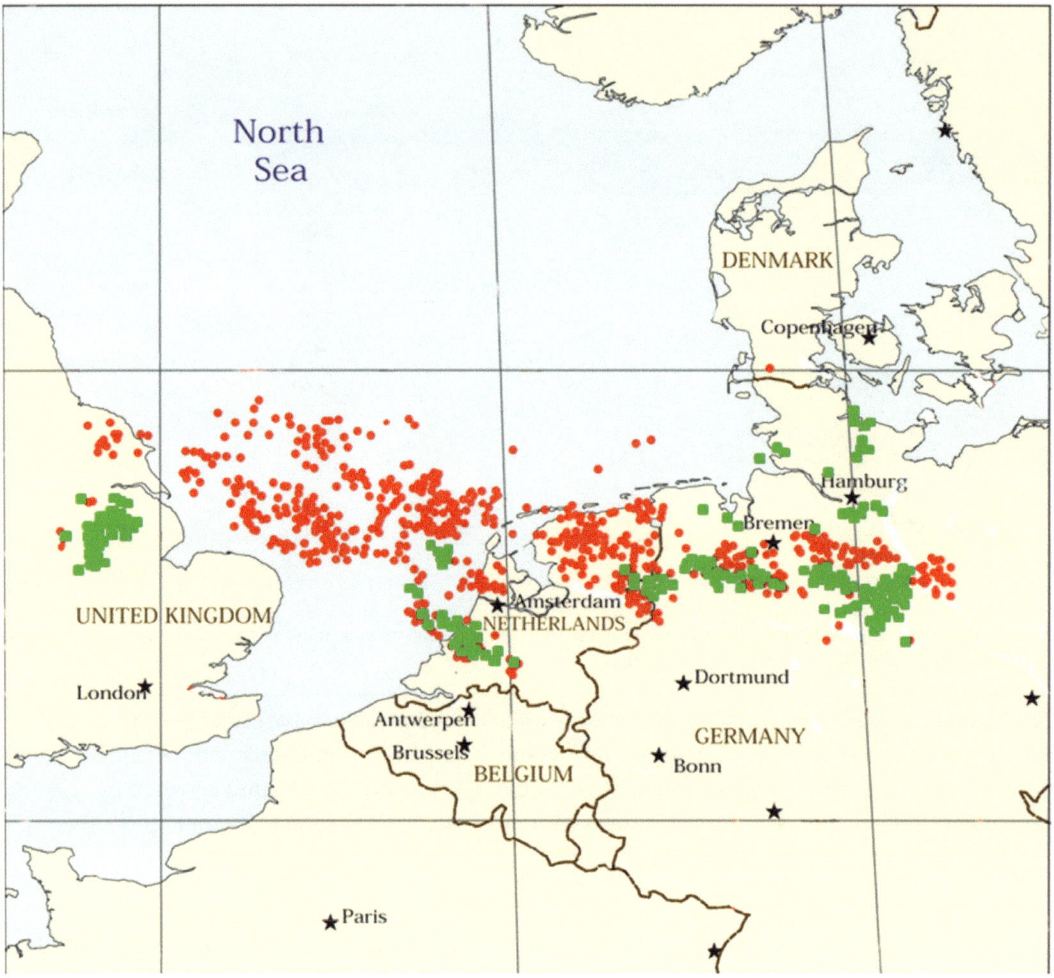

Figure 6. Oil (green squares) and gas (red dots) fields in the Rotliegend (modified from Gautier [3]).

fluid pathways. The Rotliegend is locally very hematitic and contains wide spread anhydrite and locally bitumen hydrocarbon. The alteration is probably a regional scale footprint of a deeper process (similar to that causing the Rote Fäule) that continues downward into the lower crust along a series of deeply penetrating basement faults.

2.2. Weissliegend injectite/Extrudite facies

The Weissliegend white sandstone underlies the black Kupferschiefer shale in places and is 0.3–42 m thick in the mining areas [3]. The angular and unfrosted nature of the Weissliegend quartz grains in thin section (seen in Figure 3a of [4]) indicates the sand was not formed in an eolian environment. Silica dikes and sills in the Weissliegend in the United Kingdom North Sea (photographs in Figures 8 and 14 in [5]) and the Netherlands (Figures 9 and 11 in [6]) show textures that are very similar to quartz sand injectites, such as those in the Panoche Hills and Panther Beach, California [7].

The underlying Weissliegend is interpreted to be a sand extrudite or injectite, silica mud that locally contains a matrix of copper sulfide that could have been emplaced as mud slurry composed of chalcocite and silica. For example, the Rudna Mine, in the greater Lubin district in southwest Poland, has turned out to be the largest copper mine in the Kupferschiefer (and in Europe) with 513 million tons of ore grading 1.78% copper and 42 g/metric tons silver [2]. The mineralization in the Rudna Mine is not hosted in Kupferschiefer shale, but rather is hosted almost entirely in the Weissliegend. This mine alone has more copper than has been produced from the entire Kupferschiefer. The geometry of the Rudna ore body takes the form of a large mound-like feature composed mainly of silica with lesser amounts of illite and feldspar.

From a UDH perspective, this mound-shaped feature has the appearance of a pancake-like, silica-mud volcano that is 12 m high by 1 km diameter (see cross-section Figure 4 in [8]). The Kupferschiefer has accumulated as thin carbonaceous shale facies in depressions adjacent to the inferred silica-mud volcano(s) of the Weissliegend. The Kupferschiefer interfingers with the Boundary Dolomite (which was originally recognized by Krasoñ [9]) on the tops of the Weissliegend silica mounds (see Figure 4 in [10]). The Boundary Dolomite is a dolomite-cemented sandstone that rapidly grades downward into the main, structureless sandstone of the Weissliegend. Thus, there appears to be a direct chemical and sedimentological connection between the Weissliegend chemical extrudites and the Boundary Dolomite, which is a magnesium carbonate fractionate at the top of the Weissliegend. On the flanks of the silica-mud mounds, the Boundary Dolomite then grades laterally into the carbon-rich, reduced Kupferschiefer. In effect, the Kupferschiefer in the greater Lubin area can be regarded as a volumetrically minor facies of the Weissliegend. All of the above units are regarded as chemical facies of deep-seated mud volcanism.

The Weissliegend silica extrudite mounds and ridges appear as high areas throughout the Lubin area (see cross-section in Figure 13 in [8]) and are the main hosts to the ore bodies (especially at the Rudna Mine). In the Lubin district at least, the Weissliegend and the processes that formed it are important to the genesis of the Kupferschiefer in southwestern Poland. In this sense, the Weissliegend is part of the process that is related to the broader Zechstein-Kupferschiefer mud-volcanic sequence. This interpretation is in agreement with the stratigraphic column presented in Figure 2 in [11] (as modified from Refs. [12, 13]).

The 'sandstone-hosted ore' can be further interpreted as a matrix-supported, high temperature, sulfide-rich, mud slurry that was extruded as a sulfide-matrix, mud slurry from an underlying vent source and accumulated into a chalcocite-silica-mud mound (Figure 3a of [4]). Furthermore, chalcopyrite-bornite, rhythmic lamanites grade upward into layers of subangular to subrounded, quartz grains with a partial chalcopyrite-, bornite-, and covellite-filled matrix, as shown in Figure 4 in [14] and Figures 2–4 in [15]. The copper sulfide, rhythmic bands are distinguished by their extremely light $\delta^{34}S$ isotopes, which yield values between −39 and −44. The sulfur isotopes consistently lighten upwards in a pattern that is consistent over five sections in the various parts of the Rudna, Polkowicze, and Lubin mines (Figure 3 in [15]).

From a UDH perspective, the laminae are interpreted as density-sorted, laminar flows, whereby the high-density sulfides were gravitationally sorted to the bottom of each rhythmite cycle. This gravitational sorting is consistent with slurry flowage and gravitational sorting within a silica-sulfide flow unit within the larger-scale, silica-sulfide mud mound, which is portrayed as an extrudite. The increasingly lightening upward pattern of the sulfur isotopes in the Weissliegend inferred extrudites is interpreted as being caused by progressive hydrogen reduction induced by hydrogen release from hydrogen sulfide gas during formation of various copper sulfides that were traveling as chloride complexes. The overall lightness of the sulfur isotopes may be explained by the light sulphurs that would be present in lizardite in the serpentine source region. Light sulphurs in oceanic lizarditic serpentine have been documented in an increasing number of cases summarized in **Table 1** of [16]. Hence, it is reasonable to suggest that lizarditic serpentinite might be a component of the ultramafic source near the Moho beneath the Lubin area.

3. Kupferschiefer

The Kupferschiefer (copper shale) consists of thinly laminated, bituminous, black mudstones, shales, marls, and carbonates. The Kupferschiefer traditionally is regarded as the basal member of the Zechstein Group and is 30–50 cm thick. In some areas, thin layers of carbonates are present under the Kupferschiefer. The lowermost, metal-rich, Kupferschiefer strata is 5–15 cm thick and contains the high-grade Cu-Ag-Au polymetallic and hydrocarbon mineralization in Germany and Poland at the southern rim of the Kupferschiefer Sea (**Figure 1**). If the Kupferschiefer is reconsidered as a facies of the Weissliegend silica sands as discussed above, the base of the Zechstein can be assigned to the lowest of the Weissliegend extrudites.

3.1. Kupferschiefer-Zechstein mineralogy

Copper mineralization in the Kupferschiefer *sensu stricto* consists of chalcocite, digenite, covellite, bornite, and chalcopyrite (**Figure 7**) associated with copper-arsenic ore consisting of tennantite and enargite. Galena and sphalerite commonly occur in the distal areas and only rarely in the Cu high-grade zones. Gold and silver occur as electrum in bornite and digenite (**Figure 8**). Pyrite

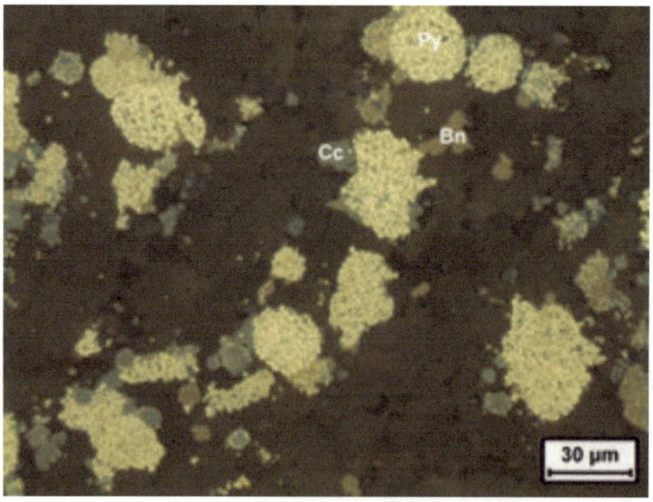

Figure 7. Microscopic mineralization in the Kupferschiefer, showing framboidal texture of copper and iron sulfides. Sample is from Spremberg drill hole 131, M 726b.4, KS T1, // ppl. Nicols.

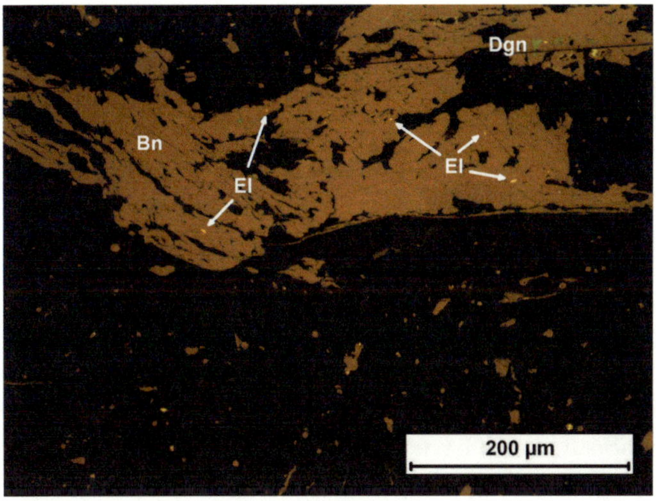

Figure 8. Massive and disseminated bornite with digenite and electrum exsolutions in kerogen- and hydrocarbon-rich Kupferschiefer shale layer T1. Sample is from Spremberg drill hole 131, M 735c, // ppl. Nicols. (Bn = Bornite, Dgn = Digenite, El = Electrum).

occurs in the Kupferschiefer and in the low-grade, mineralized zone distant from the high-grade copper mineralization. Silver and gold occur disseminated in copper sulfides as native metals and as exsolution in the form of electrum. Hematite occurs in the Rotliegend sediments and as a primary mineral in the Rote Fäule zone. Other minerals in the Kupferschiefer include marcasite, clausthalite, barite, and rutile, and also kerogen, hematite, calcite, quartz, clay minerals, and the detrital relicts of titanite, zircon, and apatite.

Copper sulfides are the main sulfides precipitated early in the mineral reaction sequence. Copper sulfides are widespread and are the major source of the economic extraction of copper. Digenite occurs as the major copper mineral. Chalcocite [Cu_2S], digenite [Cu_9S_5], djurleite [$Cu_{31}S_{16}$], and anilite [Cu_7S_4] are the copper-rich members of a mineral-series from covellite [CuS] to chalcocite. Copper sulfide deposition is important in the process of liberating hydrogen.

In the UDH model, the brine plume, which contained copper, sulfur, iron, potassium, and chloride brines, quickly ascended, depressurized, cooled, and fractionated. Once the copper and sulfur had been removed from the brine as increasingly sulfur-rich sulfides, then the remaining copper was able to combine with the chloride and precipitate as atacamite [$Cu_2(OH)_3Cl$] or paratacamite [$Cu_3(Cu,Zn)(OH)_6Cl_2$] at temperatures below 29°C. Atacamite appearance is further inferred to reflect the separation of the low temperature, low-density, Na-K-Mg brines that ultimately produced the extensive salines in the upper Zechstein. This process was not a secondary, supergene, oxidation effect. Rather, the process was a very late, primary, brine separation effect that occurred at the top of the plume, where the brines emerged onto the lower temperature paleosurface.

3.2. Kupferschiefer anomalous geochemistry

Compared to average shales, Kupferschiefer samples (especially those near structurally controlled fault 'feeders' beneath the ore deposits) are highly enriched in Ag, Cd, Hg, Mo, Co, Ni,

Cr, V, Sb, U, and Cs. The clay is also enriched in pyrite, Ba, and Sr in more distal settings. The clay of the Kupferschiefer also has strong enrichments in Re, Pb, Zn, Cu, Au, and PGE elements (Figure 7 in [17]). The carbonaceous shales contain an unusual element suite that represents an end member composition for similar, worldwide, metal-bearing, carbonaceous shales (Figure 7 in [17]). The lower 7 cm contains particularly high concentrations of the exotic element chemistry, which has made the formation important for its economic value, as shown in Figure 3 in [18].

Rather than epigenetically introducing these elements into a pre-existing, detrital shale, in the UDH model it is considered more likely that the clay-sized ides and native metals (Figures 5 and 6 from [18]) were co-formed with carbonaceous, muscovitic, mud slurries resident in high temperature (approximately 350°C), mud chambers that were deep in the Kupferschiefer conduit system. The anomalous element suite in the clays is strongly similar to the element suite in ultramafic rocks that may have participated in the generation of the sulphidic muds.

The gold, PGE, and uranium occur in the Kupferschiefer in the 'transition zone' adjacent to the Rote Fäule. The noble elements and uranium are considered to have fractionated from the Rote Fäule fluids and were deposited in the redox transition (especially the PGE component). Hence, the introduction of the gold, uranium, and PGE are inferred to have been epigenetically introduced and are not syngenetic like the earlier syngenetic, copper-silver-mud slurry mineralization in the Kupferschiefer.

Evidence for a deep-basement source for metals in the Kupferschiefer was also shown by a K-means cluster analysis of trace elements in six samples from the distal pyritic zone in the Kupferschiefer collected for the 247 ± 20 Ma, Re-Os isochron [19]. The cluster analysis showed that the rhenium and osmium are very strongly correlated with elements, such as chromium, vanadium, nickel, cobalt, and iron. These metals are major metallic constituents of peridotites and their serpentinized products. The $^{187}Os/^{188}Os$ initial ratio of 0.8 ± 1.2 suggests that the serpentinization may have been partly induced by deep, crustal, hydrothermal fluids that were integrated with radiogenic $^{187}Os/^{188}Os$ ratios, which had accumulated in the deep Caledonide basement beneath the Kupferschiefer. Mikulski and Stein [43] utilized $^{187}Os/^{188}Os$ ratios that are near peridotitic/serpentinitic initial ratios of 0.2 and obtained Re-Os dates on individual minerals to obtain ages that spanned the Permo-Triassic boundary. Kerogen hydrocarbon is strongly statistically linked to the abundance of PGE, As, V, Cr, Ni, Cu, Au, and quartz overgrowths in the Polish Kupferschiefer (Figures 5 and 6 from [18]). The positive correlations are consistent with co-eval formation of metals and kerogen and indicate that the metallization has not been randomly superimposed of a pre-existing kerogenous shale.

Evidence for coincident deposition of halogen brines and copper sulfide-illite-kerogen minerals in the Kupferschiefer is shown in Figure 4 of [20]. Copper chlorides are present on diagenetic quartz overgrowths and sylvite crystals are present in inclusions in calcite within co-existing kerogen (Figures 2 and 5 from [20]). Evidence for coincident formation of halogen brines and copper sulfide-illite-kerogen deposition is also shown on Figure 3 of [20]. The identification of sylvite is important because it provides a source of potassium for illite formation. The hypogene texture of these samples allows the interpretation that the copper chloride was directly precipitated from aqueous copper chloride from a high-density, chloride brine that was probably sourced in the basement beneath the Kupferschiefer. According to Michalik

[20], 'Halite cement in the Weissliegend sandstones is relatively common, but it occurs as small, often irregular crystals'. Paragenetic determinations by Large and others [21] (in Figures 6 and 7) show that the atacamite is closely associated with late crystallizing, low temperature minerals, such as anilite, djurleite, and yarrowite. This association may represent brine separation from the hydrothermal plume after the copper sulfides finished forming.

The characteristic correlation of kerogen hydrocarbons and metals (many of which are high temperature) in black shale-hosted metal deposits is consistent with a high temperature, hydrothermal origin for the kerogen itself. The high-energy, hydrothermal origin of the kerogen is reinforced by its close, interleaved association with high temperature, crystalline illite (muscovite) that crystallizes at high temperatures and locks in the potassium-argon age clock 350°C. The presence of illite/phengitic muscovite and metals may greatly facilitate (via aluminum-, halogen-, and Lewis acid-mediation) rearrangement of hydrocarbon compounds and alkylation at interfaces between illite and kerogen. The increase in kerogen content may be related to hydrogenization of original kerogen from hydrogen released during metal sulfide deposition from hydrothermal brines and hydrogen sulfide gas.

The deep-source, mud-volcano model offers a novel explanation for the well-known, highly anomalous, hydrocarbon occurrence in the Kupferschiefer black shale. The kerogen component of the black shale may have had a deep source, ultimately in serpentinized peridotites in the lower crust near the continental Moho. A significant component of this kerogen is possibly released into the saline brines during dehydration reactions and is progressively hydrogenated during sulfide formation to point where it may have entered the oil window. The hydrogen provided for this reaction is supplied by hydrogen sulfide with the sulfur going to form sulfides from metal-chloride ions traveling in the high-density brines. The hydrogenation explains progressive changes from PAH-dominated kerogens in the lowest Kupferschiefer to more hydrogen-saturated alkane arrays in chromatograms from the higher parts of the Kupferschiefer and lower Zechstein (Figure 1 in [22]). This chromatogram pattern was originally recognized at the Konrad Mine by Püttmann and others [22] and has since been repeatedly observed at most of the major Kupferschiefer mines from Poland to Germany.

The oil-like, alkane chromatograms of the upper Kupferschiefer indicate that the hydrogenation of the kerogen entered the oil window at the end of Kupferschiefer time. Thus, the hydrothermal mud-volcanic model is also compatible with a hydrothermal oil model. The syngenetic hydrothermal generation of oil does not require conventional maturation from kerogen to oil in a preexisting black shale. Rather, the hydrothermal oil forms at the same time as the black shale forms. In the UDH model, the carbonaceous, black shale slurries were erupted on the Permo-Triassic paleosurface. The above scenario allows the possibility that this oil and gas generation event may be correlated with the widespread petroleum accumulations that have been developed in the centers of the Rotliegend basin (**Figure 6**) in the immediately underlying Rotliegend unit (especially the Weissliegend silica extrudites; see above discussion).

3.3. Unusual textures in the Kupferschiefer

The Kupferschiefer (copper shale) has unique textures that are not like normal, detrital, sedimentary textures that are typically observed in deltaic shales. In many cases, the textures resemble hot slurries that were deposited under high-energy conditions. Thin sections reveal unusual, non-shale-like, petrographic textures (**Figure 9**). Spheroidal, orbicular particulates are present within the Kupferschiefer. The UDH model agrees with observations by Kucha [23] that are consistent with brine-silicate-flocculate interactions during upward transport from deep brine chambers (Figures 1, 12, and 15 in [23]). The concentrically banded, colloform, spheroidal, ooid-like, chalcopyrite aggregate (shown in Figure 3b in [4]) probably formed as a concentric set of rings accreted around a chalcopyrite microsphere during its upward transport in a heavy metal-rich, brine slurry.

Evidence for high-energy hydrothermalism in the Kupferschiefer is shown by the arrow-like brecciation and deposition of chalcopyrite in veins into the black shale. **Figure 10** shows an immiscible chalcopyrite sulfide slurry/melt that may have injected into the soft, carbonaceous muds of the Kupferschiefer within a high-energy, near-vent, mud-volcanic facies (also see panels A and D in plate 2 of [24]). The chalcopyrite is a matrix-supported breccia with the matrix of chalcopyrite supporting the brecciated mud. The chalcopyrite appears to have been forcefully injected into the shale, brecciating it as the chalcopyrite melt was injected. Hence, textures in the Kupferschiefer resemble hot slurries that were formed under high-energy conditions (**Figure 10**, as modified from [25]) and deposited on a paleosurface under low-energy conditions.

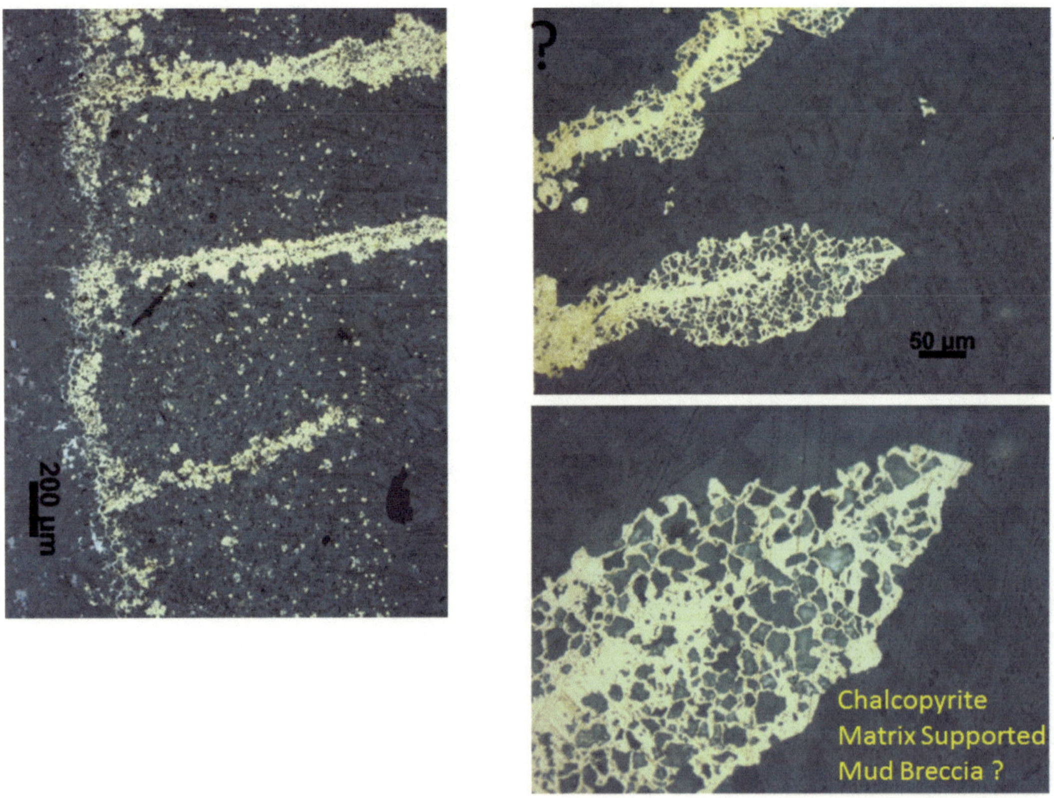

Figure 9. Spear-shaped, immiscible, chalcopyrite injectite into soft, carbonaceous muds of the Kupferschiefer. Sample is from Spremberg drill hole 131, M 734.1, KS T3, // ppl. Nicols.

http://dx.doi.org/10.5772/intechopen.72560

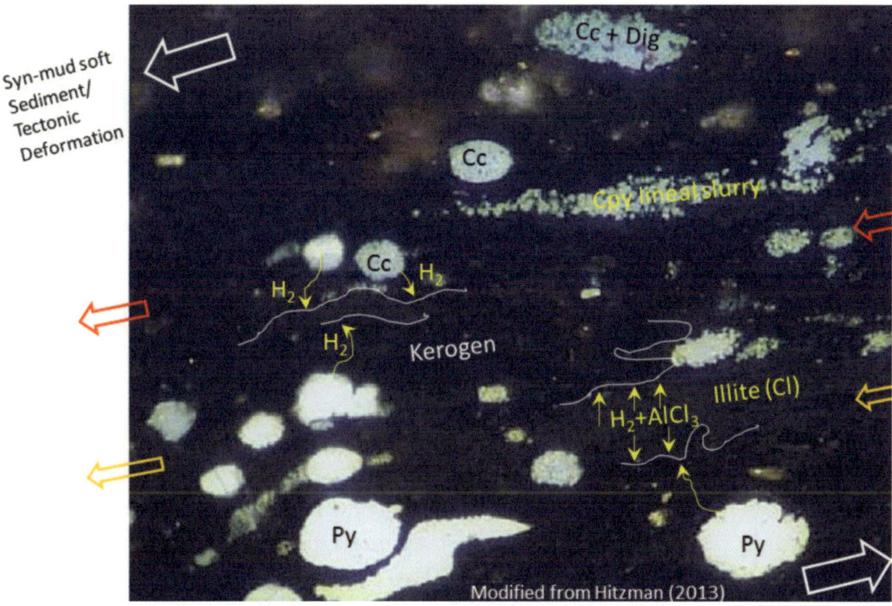

Figure 10. Interlaminated kerogen-illite carbonaceous mud with local sulphidic laminae/lineals and prominent microspherules of pyrite and chalcocite-digenite (some of which coexists with and replaces pyrite). The texture strongly suggests that the sulfides were being carried as particulates interlaminated with kerogen and illite in a sulfide-rich mud slurry.

The sulfide spherules in the copper-rich zone are interpreted as slurry droplets from sulfide suspensate in the overlying heavy brine column. Moreover, the commonly concentric, compositional laminations in the spherules (as shown in Figure 3b of [4]) suggest that the sulfides continued to grow after they were dropped into and carried in the mud-brine slurry. The sulfides are inferred to have formed as a sulphidic mud in a high-energy mud chamber beneath the Kupferschiefer shale horizon and then subsequently were erupted into the water column as high-velocity spherulitic suspensate. The sulfides were carried as lamellar slurry aggregates in soft, warm, kerogen-rich, illite-rich, carbonaceous muds, which co-formed with some of the metal sulfides. The continued growth was enabled by nearby vent-sourced, hydrogen sulfide gas and aqueous, metal-chloride flow through the soft carbonaceous mud. The framboidal (mainly in pyrite) or spherical textures of minerals in the Kupferschiefer (**Figure 7**) may document de-energization via unmixing during temperature decrease and especially during depressurization of higher energy precursor sulfide melt. The botryoidal textures, which are colloidal-like, indicate that colloidal processes can occur under both high-energy and low-energy conditions.

From the UDH point of view, the entire mud deposit that comprises the Kupferschiefer shale has an exotic character. As the high-density, brine-mud slurry ascended, it evolved from a high-energy environment of mineral stability to a low-energy environment of mineral stability. Hence, the mineralogy reflects a continuum of mineral fractionations that evolved continuously during the ascent of the mud-brine plume. In this manner, high temperature mineral slurry aggregates can reside next to low temperature mineral aggregates within the same sample at all scales. There is no need to have different high and low temperature brines affecting pre-existing detrital shales. Rather, all of the high- to low- temperature components co-evolved in the same process that began in a high-energy setting and erupted into a low-energy environment.

The previous descriptions refer to slurry textures in hand samples to microscopic samples. There is also an interesting, somewhat anomalous, larger-scale, macro-feature described by Krol and Sawlowicz as a massive sulfide vein in the basal dolomites of the Zechstein in the Lubin mine (Figures 3, 15–17 in [26]). From an ultra-deep mud volcanism approach, the feature can be interpreted as a carbonaceous mud injectite, massive sulfide, slurry sill that was injected into the dolomites of the lower Zechstein at a low angle.

3.4. High temperature to low temperature continuum of hydrothermal minerals

The apparent disparate temperature conditions indicated by various data for the Kupferschiefer can be explained by a UDH process that operated over a decreasing temperature sequence. The UDH process began in the deep crust with extremely high temperatures ranging from sulfide magmas that permissively formed near 1000°C. The UDH material then moved upward in the crust to near-surface environments with low temperature and pressure conditions, shown by the presence of anilite (which is only stable below 29°C). Ultra-deep-sourced, mud-brine, chemical volcanism can explain these disparate temperatures in a process that began in deep, high temperature regions of the lower crust and extended upward to low temperature, near-ambient conditions. At the paleosurface, the brine-mud plume extruded into the hydrosphere during the Permian-Triassic boundary.

3.4.1. High temperature suite

Spieth (in progress) conducted detailed minerographic and microprobe examinations of Kupferschiefer ore samples and prepared compositional phase diagrams. Samples were evaluated from the high-grade Kupferschiefer deposits from the Rhoen Mountains in the west, via Richelsdorf, Mansfeld, Sangerhausen, Wettelrode, Spremberg, and Weisswasser deposits in Germany to the Polish deposits in Konrad, Lubin, Polkowice-Sieroszowice, and Rudna in the east. A clear pattern of a high temperature, hydrothermal origin has emerged, as initially documented at Spremberg, Germany [27].

The high-energy, supercritical, hydrothermal plume can have melt components, such as the gold-bornite exsolution products that were formed from micro-droplets of copper-gold sulfide melt carried as a suspensate in the hydrothermal, volatile plume. **Figure 9** is a labeled photomicrograph of Kupferschiefer ores that shows interlaminated, kerogen-illite, carbonaceous mud with local sulphidic laminae/linears and prominent microspherules of pyrite and chalcocite-digenite, some of which coexist with and replace pyrite spherules.

The presence of copper selenides and high temperature digenite and djurleite in the Spremberg-Graustein deposit in Germany indicate the high temperature of formation (145–557°C) of the Kupferschiefer and associated Rote Fäule deposits [27]. Phase relations in the Cu-S-Se system also indicate the high temperature origin of the copper mineralization and the rapid cooling and exsolution within the copper minerals that also probably include gold exsolutions in clausthalite (PbSe) (**Figure 11**). Base metal-bearing selenides (klockmannite [CuSe] and krutaite [$CuSe_2$]) coexist with covellite and digenite (**Figure 12**). Experimentally determined phase relationships [28], coupled with empirical compositional data obtained by Kopp and Spieth [27] presented on Cu-Se-S ternary diagrams (**Figure 13**) demonstrate that the assemblage of covellite, umangite [Cu_3Se_2], klockmannite, and krutaite formed at high temperatures (384–343°C) [27]. The high temperatures are also indicated by the presence of high digenite [Cu_9S_5], together with djurleite. Microprobe evidence showed the copper selenides are associated with high digenite in the Rote Fäule zone of the Spremberg-Graustein deposit. These minerals are equilibrated with high temperature, hydrothermal conditions between 343 and 384°C [27]. Stability fields of sulfide minerals present in the Kupferschiefer range between 72 and 557°C and up to 1120°C for digenite (high) (**Table 1**). Sources of information were included in [29–37].

Microscopic textures and compositions of the sulfides are consistent with the presence of a high digenite-high chalcocite-bornite sulfide melt that existed at temperatures above the supercritical-subcritical boundary for water. These sulfide phases might have unmixed as

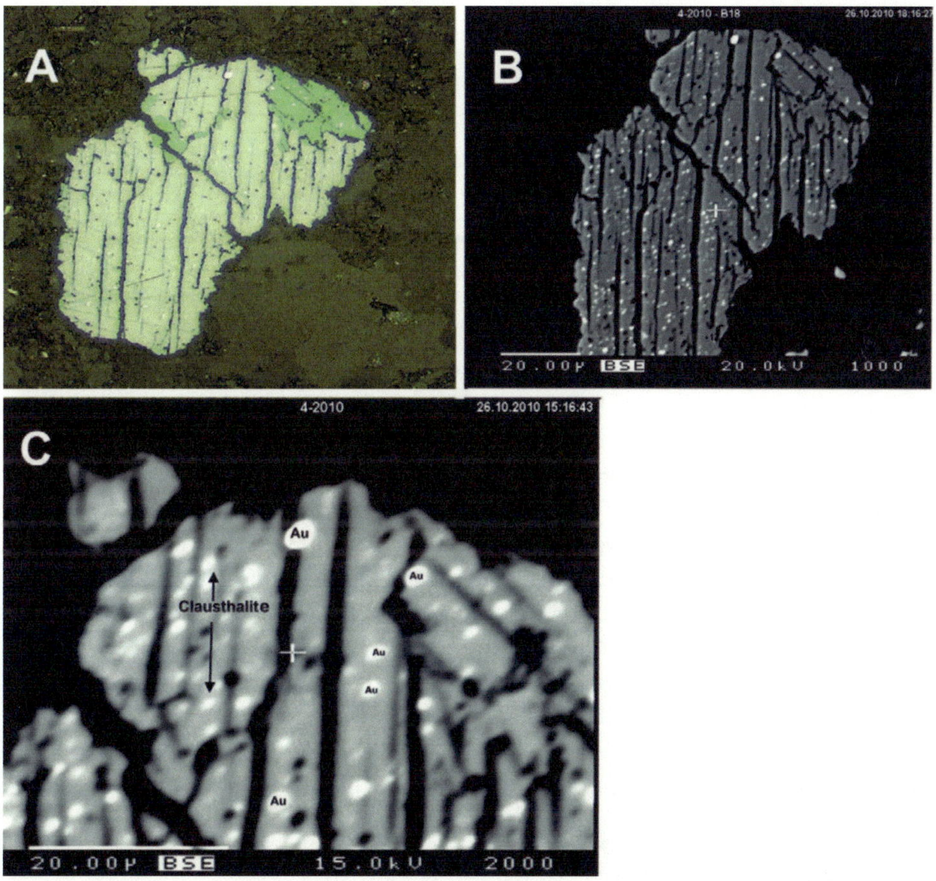

Figure 11. Digenite from the Rote Fäule at Spremberg: A. Reflected light × 1000, // polars. B. The BSE image reveals numerous precipitates of native gold and clausthalite. C. The image shows the enlarged upper part of the image from the top right with rich gold and clausthalite content.

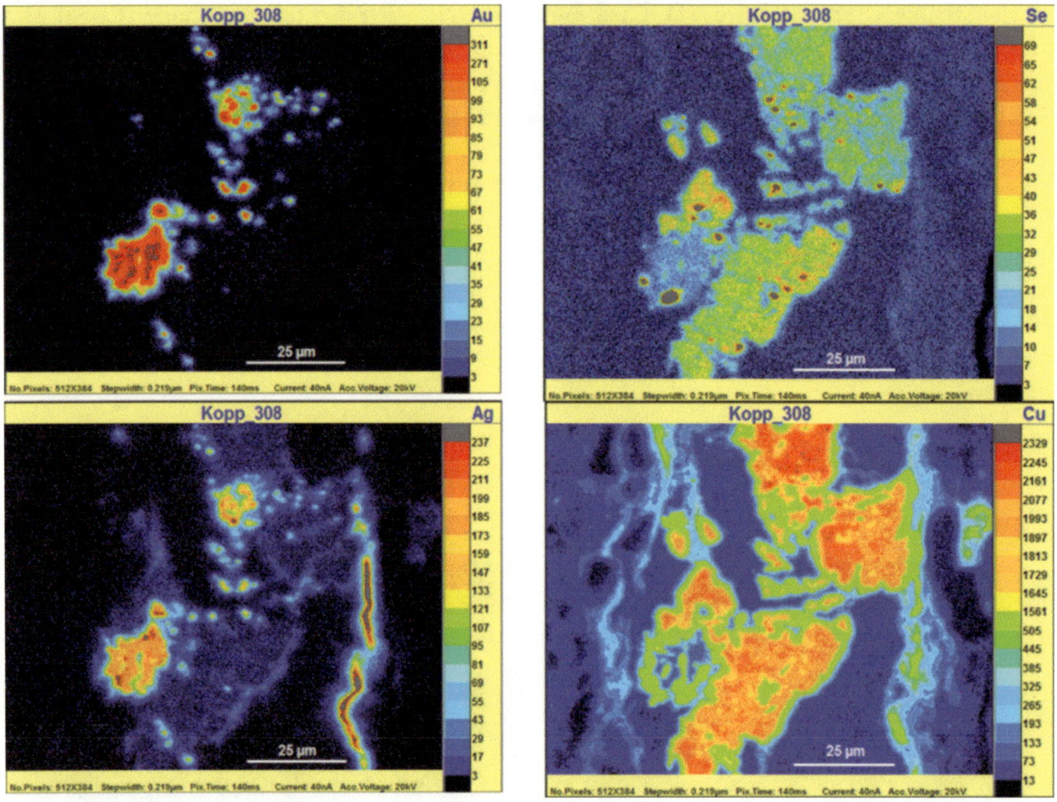

Figure 12. Ag-Au-Se-Cu-S distribution in a copper sulfide grain with electrum and klockmannite included in a chalcocite matrix from the transition zone between Rote Fäule and Kupferschiefer at Spremberg.

immiscible phases of high temperature, copper sulfide melts that initially existed as nano-sized, sulfide droplet emulsates carried in high temperature, supercritical brines. The brine-mud-kerogen mix is inferred to have been partly sourced from deep, steatitic, brine chambers that were derived from the dehydration of serpentinite sources in the deep crust. The brines were then carried upward to the eruptive site at the Kupferschiefer paleosurface. As the brines ascended, droplets of similar composition and density may have aggregated into larger droplets that comprise much of the spherical and orbicular-shaped, sulfide grains. When the brine-sulfide magma crossed into the subcritical phase for saline water below 400°C, the sulfide droplets ionized and crystallized into fibrous forms and other cryptic forms. When temperatures dropped below 500°C (and especially below 400°C) and the brines ionized, covellite began to exsolve from high digenite or high chalcocite and exsolution laminae of covellite formed within chalcocite/digenite. Ionization of the kerogen-bearing brine component below 400°C induced a rapid hydrogenation of the kerogen hydrocarbon component and may have led to the production of a hydrothermal oil component between about 370 and 320°C.

The presence of phengitic muscovite (which is frequently referred to in the literature as illite) can also be used as evidence for high temperatures of formation. The presence of high purity, high crystallinity phengitic muscovite, which texturally coexists with high temperature sulfides, also indicates a temperature of at least 200°C. Even higher temperatures are indicated by

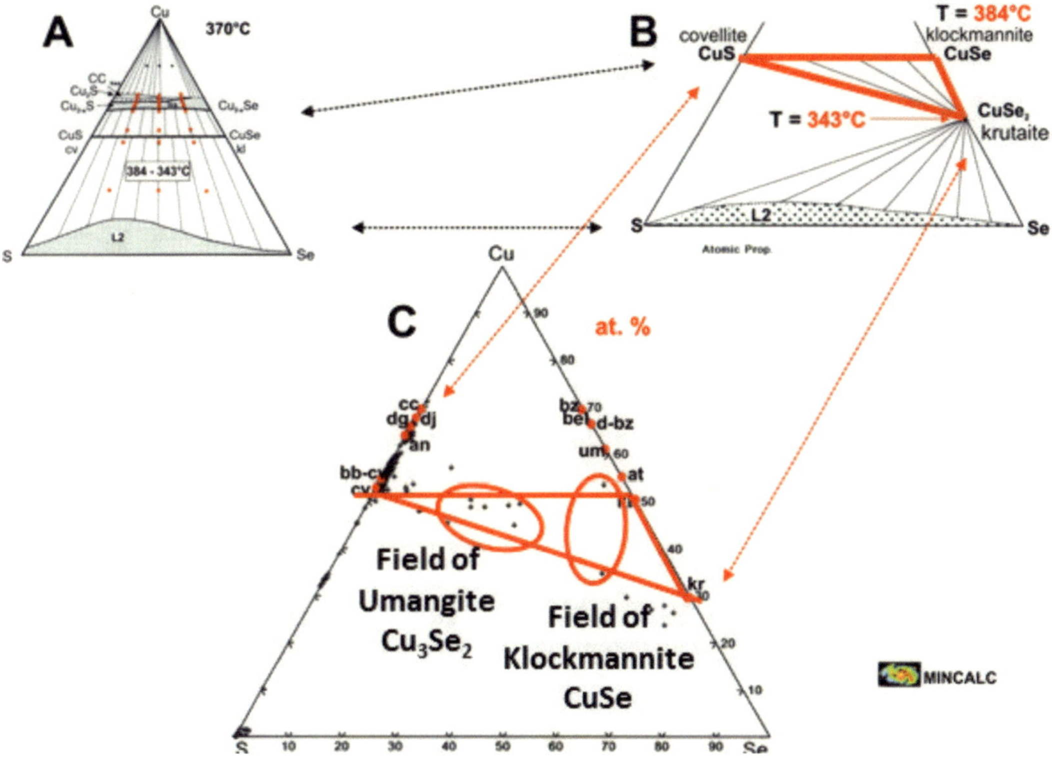

Figure 13. Ternary diagrams showing phase relations in the Cu-S-Se system between ~384 and ~323°C (modified after [28]). Red solid dots in A represent the composition of the Bernardini experiment [28]. Red triangle in B is the compositional field for the krutaite-klockmannite-covellite system. C shows composition of Cu-S-Se analyses of copper-silver deposit at Spremberg (modified from Kopp et al. [27]).

Cu-sulfide	Composition	System	Stability	References
Chalcocite (Low)	$Cu_{1.99–2}S$	Monoclinic	$T = 103°C$	1, 2
Chalcocite (high)	$Cu_{1.98–2}S$	Hexagonal	$T = 103–435°C$	2
Chalcocite (high-P)	Cu_2S	Tetragonal	$P > 1$ kbar, $T = 500°C$	3, 2
Djurleiite	$Cu_{1.97}S$	Orthorhombic	$T = 93°C$	1, 2
Digenite (low)	$Cu_{1.75–1.8}S_5$	Cubic	$T = 76–83°C$	5, 2
Digenite (high)	$Cu_{1.73–2}S_5$	Cubic	$T = 83–1129°C$	6, 2
Anilite	$Cu_{1.75}S$	Orthorhombic	$T = 72°C$	7, 4
Yarrowite	Cu_9S_8	Hexagonal	$T = 157°C$	2
Bornite	Cu_5FeS_4	Tetragonal	$T = 228°C$	8, 2
Chalcopyrite	$CuFeS_2$	Cubic	$T = 557°C$	8, 2, 9

References: 1. Roseboom [29]; 2. Vaughan [30]; 3. Skinner [31]; 4. Potter II [32]; 5. Morimoto and Koto [33]; 6. Morimoto and Kullerud [34]; 7. Morimoto et al. [35]; 8. Schröcke [36]; 9. Barton [37].

Table 1. Temperature and pressure stabilities of various forms of copper sulfide minerals in the Kupferschiefer [27].

dated, high-quality illite/muscovite separates that probably locked in their K-Ar ages near 350°C, based on temperature-closure data for high-quality muscovite near 252.5 Ma. The 252.5 Ma date probably also marks the changeover from high temperature sulfide deposition to lower temperature, sulfide deposition. Re-Os dates on individual bornite and chalcocite samples (which include the main volume of the high temperature suite) are all older than 252.5 Ma.

The temperatures of selenide formation are in the same temperature range as those in the hydrothermal hydrolysis experiments reported by Lewan and others [10] that created oil from Kupferschiefer rocks. In a 1978 experiment, Lewan [38] created oil in the laboratory by submerging crushed Kupferschiefer rocks in a 1-l container of water and heating it to temperatures between 300 and 360°C (greenschist metamorphic conditions) where after 72 hours it became temperature-equilibrated. At the end of the experiment, oil was floating on top of the water. No oil was formed at temperatures below 300°C or without the presence of water. In effect, Lewan created hydrothermal oil. Most importantly, alkane oils were created in the same high temperature ranges as the stability of the Cu-Se sulfides mentioned above. The Kupferschiefer pyrolysis experiments represent the first pyrolysis experiments done in the presence of a natural hydrothermal metal system. Hence, the Kupferschiefer system represents the first case history where well-constrained temperatures for sulfide formation match temperatures at which Lewan created hydrothermal oil by hydrous pyrolysis. Oil synthesis probably formed in the Kupferschiefer under high temperature conditions at the same time and similar temperatures as were present when the metal sulfides formed.

3.4.2. *Low temperature suite*

In the near-surface environment of the Weissliegend-Kupferschiefer chemical sand-mud pulse, temperatures in the main Kupferschiefer dropped below 350°C, allowing the low temperature assemblage to precipitate from the metalliferous, K-Mg-Na-chloride brines. At this point, orange bornite (>268°C), spionkopite [$Cu_{39}S_{28}$], geerite [Cu_8S_5], yarrowite ([Cu_9S_8] also called baubleibender covellite, stable below 157°C), djurleite (stable up to 100°C), and anilite (not stable above 39°C) formed. When temperatures dropped below 100°C, chalcocite may have converted into djurleite (**Figure 14**), which marked the beginning of the end of the sulfide formation process, which was completed when anilite formed below 39°C. At this point, the brines had probably erupted/flowed onto the sea bed, spewing copious amounts of metalliferous muds in low-relief, pancake-shaped, mud volcanoes. When the textural data discussed above are combined with the temperature data and exotic mineral data in the overlying Zechstein salines, they provide powerful evidence for a mud-brine, chemical volcanic model that is deep-sourced.

Under extremely high copper concentrations in the chloride brines, atacamite may have been precipitated as microspheres near covellite-muscovite/phengite interfaces. The appearance of atacamite may have signified the separation of the low-density, K-Mg-Na-Ca-chloride brine at the end of covellite deposition. This low-density brine would then separate from the higher density, sulphidic and carbonate brine components just below the paleosurface and rapidly form the extensive Zechstein brine deposits, which include saline facies. These saline formations would contain inclusions of exotic minerals, such as talc, serpentine, and clinochlore that crystallized deep in the brine system in the lower crust, probably at temperatures well above 350°C.

http://dx.doi.org/10.5772/intechopen.72560

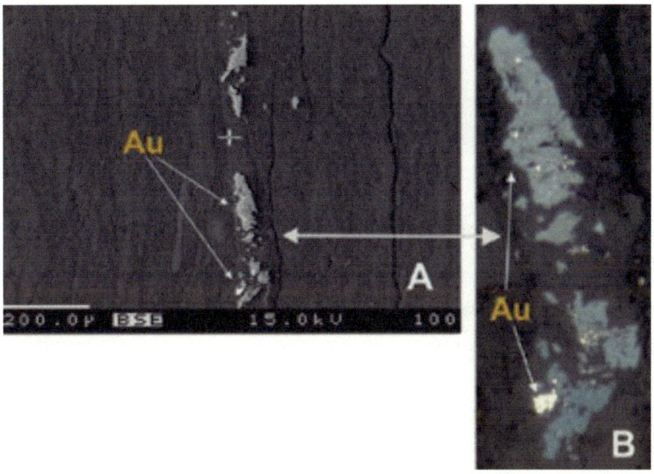

Figure 14. BSE image showing discrete grains of copper sulfides containing native gold inclusions from the lower Kupferschiefer that are interlaminated with kerogen-illite shale; B. Reflected light, X500, plane polars showing chalcocite, light gray djurleite, gold exsolutions, and small grayish inclusions representing native gold, electrum, or clausthalite from the transition zone between Rote Fäule and Kupferschiefer at Spremberg.

3.5. Zoning in the Kupferschiefer-Zechstein

The Kupferschiefer-Zechstein is a multiply zoned, hydrothermal, brine-mud system. The Kupferschiefer deposits are zoned on small and large scales, both laterally and vertically. On a mine scale, numerous studies have shown the Kupferschiefer shales are zoned vertically within the 1-m thick section. Vertical depositional zoning from the bottom to the top of the Kupferschiefer is: (1) copper sulfides (chalcocite → digenite → chalcocite-digenite-bornite → chalcocite-digenite-covelite → digenite-covelite-chalcopyrite → chalcopyrite-bornite-galena) → (2) upward into the Zechstein carbonate marls of galena with minor chalcopyrite → sphalerite → pyrite.

On a lithostratigraphic basis at the mine to district scale, the Kupferschiefer-Zechstein is characterized by enriched polymetallic mineralization in four zones from top to bottom (**Figure 5**): (1) hanging wall Zechstein carbonate, (2) Kupferschiefer shale carbonaceous main horizon, (3) footwall Weissliegend silica sands, and (4) Rote Fäule, which post-dates the previous three rock types as a typically replacive, hydrothermal alteration. The hanging wall mineralization is in the Werra carbonate rocks of the Zechstein Group dolo-limestone and is 1–10 m thick. The main mineralized horizon is in Kupferschiefer black shale and is less than 0.5–2 m thick. The foot wall mineralization is in the Weissliegend silica sand unit and is 1–15 m thick. The highest copper grades, which strongly focused historical mining activity, occurred at the base of the Kupferschiefer shale. Copper grades typically ranged between 8 and 20% in a carbonaceous layer about 2–6 cm thick.

Within the mine areas, the Rote Fäule independently overprints the above-described mineralization and consistently contains elevated gold, which reaches ore grade in a transition zone between the pre-existing Kupferschiefer-Zechstein mineralization described above. PGE enrichment displays a preference for the transition zone, which indicates their deposition is associated with the shift from the oxidized Rote Fäule source of the PGE-U to the reduced

Cu-Ag (Re) Kupferschiefer shales. The reduction effect is thought to have induced deposition of the PGE suite.

At a regional scale, lateral zoning established mineralization belts are oriented concentrically outward from the Rote Fäule (used a frame of reference only) in the following sequence: hematite → gold →PGE → thucholite → chalcocite → bornite → chalcopyrite → galena → sphalerite → pyrite (**Figure 15**). A general lateral succession (which is sub-horizontal, tabular, and belt-like) of the sulfides begins with copper, which is also rich in silver and which occurs in the transition zone close to the border of the hematitic, Rote Fäule alteration zone. Laterally, farther outward of the copper zone is the lead and zinc zone; even farther outward is the pyrite-rich zinc zone. Lateral zoning occurs over thousands of square km on the scale of the entire Polish Basin. The zonation of hematite and copper in the southwest part to zinc and pyrite in the northeast indicates the direction of paleo-fluid flow. In the mud-volcanic model, the paleo-fluid could include nano- to micro-sized flocculates and suspensates. The high-grade, copper-silver ores of commercial interest are typically located in a sub-horizontal, tabular, belt-like bodies that are 1.5–3.5 km wide and can be more than 10 km in length near the boundary of the Rote Fäule area (e.g. at Spremberg in southeast Brandenburg and Lower Silesia [27]).

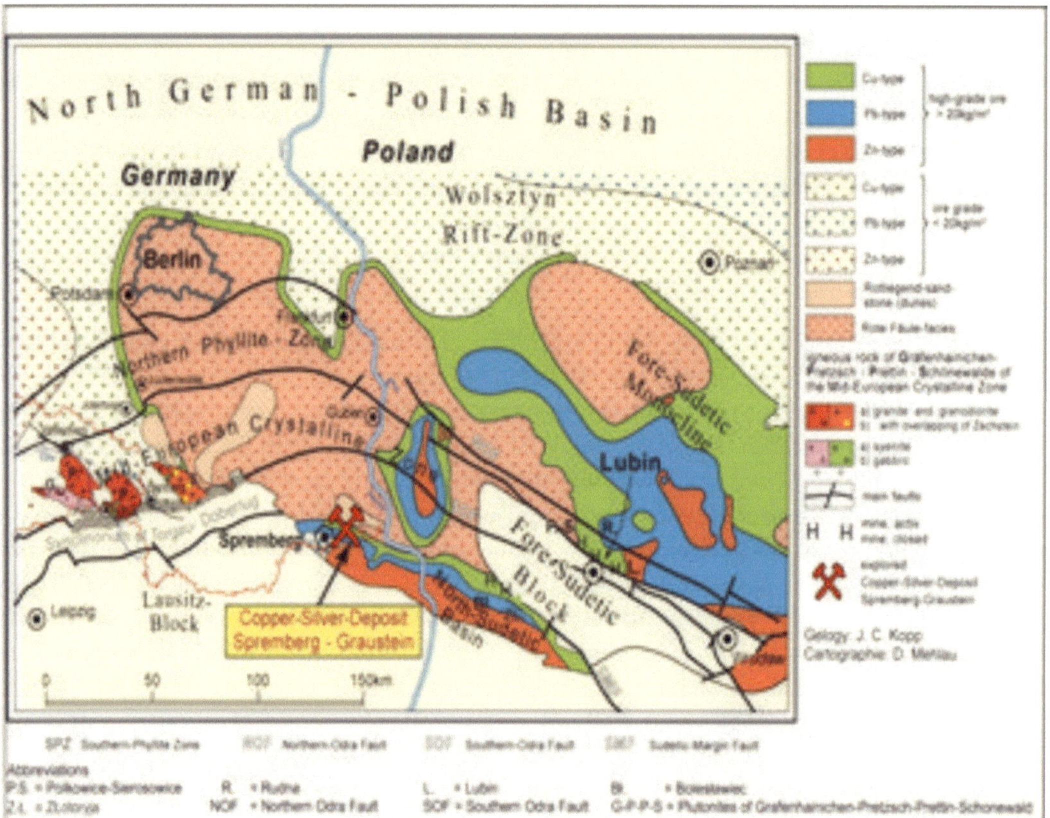

Figure 15. Basin-scale, lateral zoning of Kupferschiefer metallization.

Gas and minor oil fields are hosted in Weissliegend sandstones and basal Zechstein carbonates. Gas is locally co-extensive with the copper deposits in the zinc zone at the mine scale. Gas is low in H_2S and high in N_2 and He, which is similar to chemical signatures of kerogen in the mantle. On a basin scale, oil occurs near the centre and north of the gas occurrences (**Figure 6**). In southern Germany and Poland, an oil generation zone may occur in the outer Zn zone. The regions with gas fields coincide with copper regions. Gas resources are locally world class, as in the Groningen field of northern Netherlands [3].

3.6. Paragenesis and age of the Kupferschiefer minerals

The paragenesis of ore minerals in the Kupferschiefer in 10 metalliferous associations is shown in **Figure 16**. The sequence of ore minerals in the Kupferschiefer is modified from Jung and Knitzschke [39]. The paragenesis appears laterally and stratigraphically as: the copper-enriched zone at lower stratigraphic levels, with lead and zinc best developed in upper layers, and with pyrite-type metallization appearing in the uppermost or distal facies. Similar sequencing appears in the vein style, cross-cutting metallization.

The Kupferschiefer-Zechstein sequence at Spremberg shows strong chemical and paragenetic relationships between sulfides, kerogenous shales, illite, saline minerals, dolomitic carbonates, silica, and calcite (**Figure 16**) [27] as modified after [40, 41]. The reaction sequence is: (1) early silica; (2) with copper-silver-rich, illitic, carbonaceous shale becoming more dolomitic and

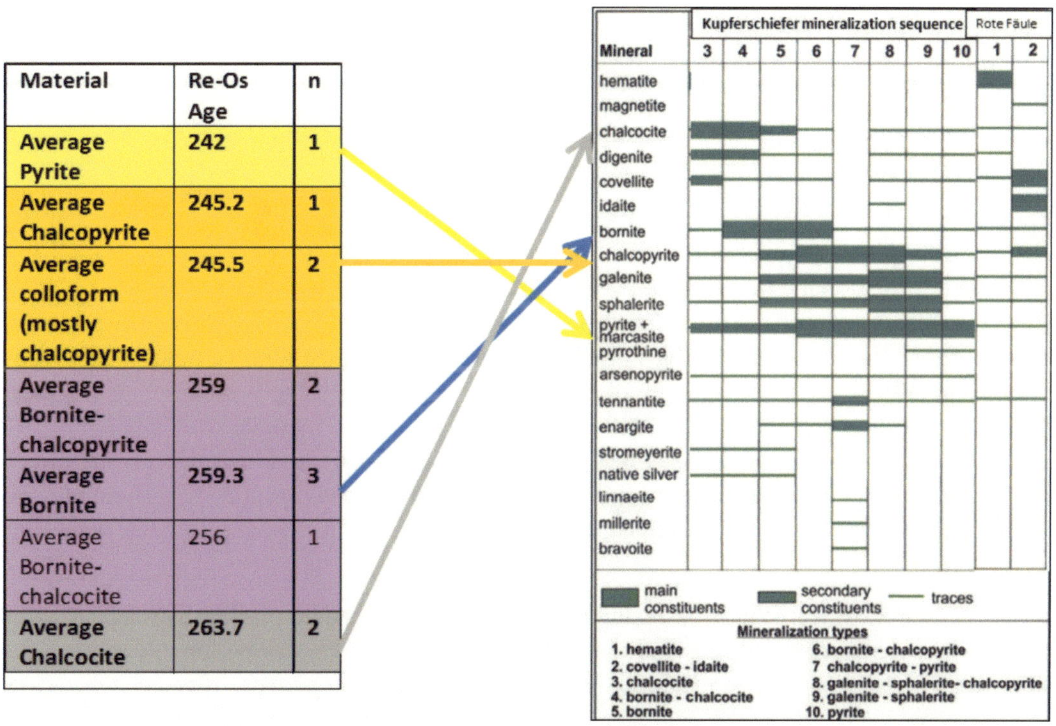

Figure 16. Re-Os ages on sulfides compared with modified paragenetic sequence for the Kupferschiefer (ages from Refs. [4, 43]; modified paragenesis is from Refs. [27, 40, 41]). Stage 2 in the right column also contains major Au-(Pt-Pd-U-Co) and is equivalent to the SOS mineralization (described in Refs. [11, 13]).

bitumen-rich up-section; (3) then an upper, zinc-rich zone associated with dolomitic carbonate; (4) followed by calcitic carbonate; (5) with saline chemical sediments forming a lithocap above the carbonates; and (6) the Rote Fäule representing late stage, oxidized, hematitic alteration that post-dates marly Werra dolomitic carbonate and post-dates underlying Kupferschiefer and Weissliegend silica sand units.

The sequential deposition of silica-rich strata is followed by carbonate-rich strata. Sometimes, biologic activity utilizes the silica-rich effluent in the proximally located flux of hydrocarbons. Oil generation is coincident with steps 8 and 9 in **Figure 16** as the oil is generated and deposited along with the galena-sphalerite-chalcopyrite mineralization.

The paragenesis shown on **Figure 16** is augmented by observations pertaining to the non-sulfide mineralogy, which is also shown as a Kupferschiefer rock section in **Figure 3**. The early chalcocite-digenite mineralization is typically accompanied by silica (which occurs as fragmental, silica, extrudite and injectite bodies) that comprises the Weissliegend beneath the Kupferschiefer. Silica is maximized in unit 3 (the Weissliegend). Silica rapidly decreases upward across a sharp contact with the Kupferschiefer copper shales, where illite/muscovite rapidly increases along with kerogen hydrocarbon, both of which is present in the Weissliegend, but only in minor amounts. The illite and kerogen hydrocarbon are especially abundant in the basal Kupferschiefer, especially in unit 2 and to a lesser extent in unit 3. A rapid gradation occurs upward into the Kupferschiefer, as marked by the appearance of dolomitic marls, which progressively become more abundant than the illite. The dolomitic marly unit coincides with units 6, 7, and 8 in **Figure 16**. The dolomitic marls are succeeded upward by dolomite and subsequently upward by calcitic carbonate. The dolomite coincides with the appearance of abundant sphalerite in unit 8 and especially in unit 9. Pyrite appears in unit 6, but its occurrence as the sole sulfide coincides with the presence of calcitic carbonate in unit 10. The Permo-Triassic boundary approximately corresponds with unit 6.

There is a shift from the chalcocite-digenite-covellite assemblage of stages 3, 4, and 5 to the chalcopyrite-sphalerite-galena-dominated assemblage of stages 7 through 9. A prominent age gap is also present in the Re-Os data (**Figure 17**) that may suggest two stages of Kupferschiefer copper sulfide mineralization: the first being copper-sulfur dominated and the second being copper-iron-sulfur dominated. The ages from the higher quality illite samples taken from **Table 2** in [42] yield an average age of 252.5 ± 4.5 Ma for eight high-quality illite samples. These samples generally were pure (>6.9 wt.% K_2O), were of intermediate grain size (<2 or 2–1 μm samples that tend to have the purest [highest K_2O contents]), and had high crystallinities (<48% 2 M disordered component [or more than 52% 1 M ordered component]). There may be a time break of about 10 Ma between the carbonaceous Kupferschiefer and the overlying Zechstein carbonates.

In the UDH mud-volcanic model, the illite ages reported in [42] date primary crystallization of muscovitic mica during the mud-volcanic process. The highly crystalline, high-quality 'illite' samples selected from [42] are inferred to represent primary muscovite that crystallized in mud-volcanic vents, where the potassium-argon was locked in at relatively high temperatures (around 350°C). The progressively older ages in increasingly more smectitic illitic clays are interpreted to represent the incorporation of extraneous, excess argon in the more distal pyritic zones on the Kupferschiefer shale time line (see Figure 4 in [42]). Thus, K-Ar dates do not record

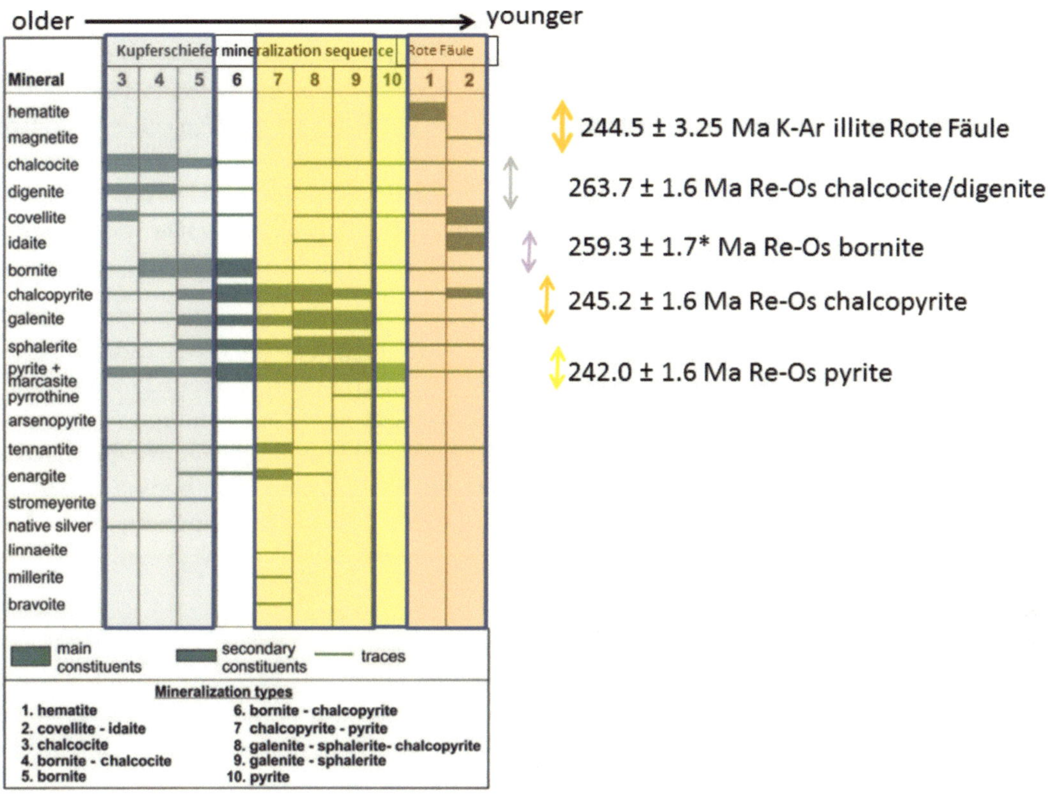

Figure 17. Summary of isotopic age calibration for main stage Kupferschiefer and overprinted Rote Fäule.

diagenetic effects in a pre-existing detrital shale, or do they date secondary, neoformation, or recrystallization of illite by hydrothermalism related to Kupferschiefer mineralization. The 252.5 Ma age is considered an average age for the primary emplacement of the greater, more generic, Zechstein-Kupferschiefer metal system in the Lubin area. The error bars may reflect metal deposition at the early part (circa 257 Ma), which would be the chalcocite-digenite sulfides in the Weissliegend and basal Kupferschiefer. None of the illite ages come from the underlying Weissliegend unit. The younger end of the age range may reflect iron-copper sulfide deposition in the upper Kupferschiefer and lower Zechstein Werra dolomite sequence. Three of the ages collected specifically from the Rote Fäule yield an average age of 244.3 $\pm$ 3.25 Ma, which is consistent with the observation above that the Rote Fäule is overprinted on the earlier Kupferschiefer rocks.

A second set of 12 rhenium-osmium (Re-Os) ages is available in [4, 43] that more precisely constrain aspects of the metallization in the main Kupferschiefer. These ages were obtained from more targeted, individual, sulfide samples that provide ages for specific sulfide mineralization events. Obtaining Re-Os ages from specific mineral assemblages provide more specificity than Re-Os isochrons, which assume that the samples on a given isochron represent the age of deposition. The high-energy, mud-brine, chemical volcanism model presented here proposes that the sulfide grains are exotic and are not necessarily equilibrated with the same Re-Os reservoir as their contained kerogen matrix.

Thus, the high-quality ages allow a reconstruction of specific events in the sequence of events for deposition of the Zechstein-Kupferschiefer system. The illite ages allow separation of the main Kupferschiefer event at 252.5 Ma from the overprinting Rote Fäule event at 244.5 Ma. The Re-Os data allow a separation of the sulfide deposition events in the Kupferschiefer into two main stages. A copper-sulfur-dominated event (chalcocite-digenite-covellite) affected the Weissliegend and main Kupferschiefer unit from 267.7 to 259.3 Ma (**Figure 17**). A second event of iron-copper-sulfur-dominated metallization mainly affected the carbonate marls of the Lower Zechstein and ranges in age between 245.2 and 240.6 Ma based on averaged data for chalcopyrite and pyrite samples.

The entire Zechstein-Kupferschiefer event from the Weissliegend to the lower carbonate section (Werra) of the Zechstein ranges from 264 to 240 Ma, a time span of about 25 Ma. Late in the sequence, the Rote Fäule appears at approximately 245 Ma. Consideration of the geographic distribution of the illite ages shows that the Zechstein-Kupferschiefer event affected a much larger area than the metal-rich areas around Lubin, Konrad, and other mining areas. The long-lived process affected the entire Zechstein-Kupferschiefer throughout the area that encompasses the Polish Zechstein basin comprising the eastern part of the overall Zechstein basin in north central Europe (**Figure 1**).

Significantly, the age of the Zechstein-Kupferschiefer spans the Permo-Triassic boundary, based on the combined Re-Os and illite ages. The boundary between the Permian and the Triassic can be placed at the top of the black shale of the Kupferschiefer and below the Zechstein carbonates (**Figure 3**). The current Permo-Triassic boundary is radiometrically calibrated at 252.2–252.5 Ma, which is the same as the age of the high-quality illite data for the Kupferschiefer. This boundary also falls near the midpoint of the apparent age gap between 245.5 and 259 Ma in the Re-Os mineral dates (**Figure 16**).

4. Zechstein Group

The Zechstein Group is divided into four cycles (**Figure 2**), each of which consists of marls and limestone at the base, overlain by layers of salines, such as anhydrite and halite [2, 44]. Magnesium and potassium chlorides and sulfates (e.g., sylvite [KCl], polyhalite [$K_2Ca_2Mg(SO_4)_4{\cdot}2H_2O$], kieserite [$K_2Ca_2Mg(SO_4)_4{\cdot}H_2O$], and tachyhydrite [$CaMg_2Cl_6{\cdot}12H_2O$]) are well developed in cycles 3 and 5 of the Zechstein Group.

In addition to the well-known saline minerals above, exotic, highly magnesium minerals (such as talc, clinochlore, and serpentine) typically occur in the lower part of the saline sequence in the Zechstein saline rocks that overlie the Kupferschiefer (**Figure 1** and **Table 2**). The above minerals occur in refractory, shaly interlayers, and residues within saline units. The occurrence of serpentine in insoluble clays in the Emsland occurrence may represent the direct emplacement of serpentine-sourced muds as extrudite slurries into the Lower Zechstein, Werra anhydrites. At Emsland, the serpentine is partly present as ooids, which may have been pelletized during upward transport in a conduit system.

Talc occurrences (which are the most frequent of the unusual mineral occurrences), as well as clinochlore and especially serpentine occurrences, indicate the mud-brines were formed in an

Location	Mineral association	Age	References
Zechstein-Exdale, Yorkshire, England	Mainly halite and anhydrite (85%); magnesite, **talc on cleavages in halite, talc in anhydrite-magnesite-halite-talc-rock,** halite, minor quartz in seams in anhydrite; halite or sylvinitic mudstone, halite and anhydrite. Boracite, carnallite, hematite, magnesite, rinneite, **talc**, and **euhedral quartz** are generally minor constituents	Zechstein (Permian)	[45–47]
Zechstein Salt, Mors dome, Denmark	**Euhedral quartz** in carnallite-sylvite-halite with fluid inclusion history that ranges from near 300°C to circa 150°C during crystallization and presumed brine ascent	Zechstein (upper Permian)	[48]
Zechstein, in the Frenswegen 3 drillhole, Emsland, Germany	**Serpentine (as antigorite)** occurs with chlorite (probably **clinochlore**) and **talc** interlayered with anhydrite, halite, salts at 1960 and 2035 m depth in the Frenswegen drillhole	Werra Salt unit in the lower Zechstein	[49]
Zechstein-Gorbelen Salt Dome, Germany	Dolomitic carbonates contain hydrocarbons, halite, anhydrite crystals, quartz, pyrite, accessory minerals of quartz, celestine, sphalerite, rare potassium feldspar, fluorite, **talc**, and galena in a long-mined salt deposit	Zechstein (upper Permian) originally, remobilized in Cretaceous	[50, 51]
Konigshall-Hindenburg salt deposit, Germany	**Clinochlore** occurs associated with boracite, carnallite, celestine, dolomite, gypsum, halite, hematite, magnesite, magnetite, illite, pyrite, quartz, sylvite, and tachyhydrite in an ancient potash mine closed in 1970	Upper unit in the Zechstein (uppermost Permian)	[52]
Zechstein, West Poland	**Talc** commonly occurs with celestite in veins in anhydrite; rare quartz, Sr. anomaly in dolomite mounds	Zechstein, Upper Permian	[53]
Zechstein-Kłodawa salt dome, central Poland	Halite, carnallite, kieserite, sylvite, anhydrite, polyhalite, dolomite, chlorite and illite were identified as the main mineral components in the dome, and kainite, langbeinite, picromerite, glauberite, löweite, douglasite, tachyhydrite, bischofite, calcite, magnesite, celestine, gypsum, alunite, boracite, congolite, quartz, **talc,** pyrite, sulfur, fluorite, glauconite, **forsterite**, orthoclase and hydrotalcite-like **motukoreaite**, saponite	Zechstein (Permian)	[54]

Table 2. Locations of talc, serpentine, clinochlore and euhedral quartz associated with saline, potash, K-Na-Mg Cl brine deposits.

ultra-deep environment in an underlying steatitic reaction chamber. These minerals could have been instantaneously erupted as mud slurries into the saline deposits as mud slurries, as they are commonly associated with anhydrite at the halite-anhydrite transition. Polyhalite is also a common associate of talc.

A worldwide survey for these exotic saline components also identified talc, clinochlore, lizardite, serpentinite, and euhedral quartz occurrences from 27 locations, including 11 occurrences of unusual minerals in the Zechstein (**Table 2** and **Figure 1**). These occurrences were documented in [45–54].

4.1. Zechstein saline deposits

The hypersaline Zechstein brines have a unique, nonmarine chemistry. Bodine [55] showed that Zechstein-type salt deposits have a distinct magnesium-potassium geochemistry that shows a composition half way between the aluminum corner (Al_2O_3) and the $KAlO_3$ corner (Figure 10 in [55]). The main compositional field is completely displaced from the normal marine composition, which is more aluminous [55]. Significantly, the illite clay components in the insoluble compounds are typically much more potassic (>0.85 molar potassium) than in the more detrital marine counterparts.

The illites in the insoluble fraction of the Zechstein-Kupferschiefer are also highly crystalline and contain only very minor smectitic interlayers compared to their detrital counterparts [55]. The high crystallinity characteristics of the Zechstein-Kupferschiefer illite are more typical of hydrothermal illite. Such illites could have been made originally under hydrothermal conditions in a conduit system and later erupted as crystalline illite floculates into the saline brine, surface environment. In the case of the Zechstein, the setting may have been a playa lake or shallow marine environment, but not a deep sea environment, as suggested by much of the Zechstein-Kupferschiefer literature. Rather than appeal to rapidly changing water depths, it is more likely that the shales were also erupted as carbonaceous extrudite muds erupted into shallow marine waters from high temperature, fissure vents.

The presence of talc, clinochlore, euhedral quartz, and serpentine probably record a higher temperature history resident in deep mud-brines prior to their ascension in hydrothermal mud-brine plumes and prior to their eruption/flow onto the paleosurface into what are construed as shallow lake/marine muds. Serpentine, talc, and clinochlore, generally, form at temperatures above 300°C, which at 25°C/km indicates a minimum of 11 km depth below the paleosurface. In addition, a euhedral quartz crystal from the Mors Salt dome in northern Denmark studied by Fabricius [56] contains fluid inclusions that equilibrated with polyhalite at homogenization temperatures up to 180°C (Figure 4 in [56]). The crystal nucleated on an early carnallite [$KMgCl_3{\cdot}6H_2O$] clot at greater than 180°C and trapped 80°C fluid inclusions during its late growth history as it ascended from depths of circa 3 km.

On a global basis, the Zechstein salines are not a unique occurrence [57]. They occur throughout geologic time (Figure 13 in [57]) and seem to be associated with break ups of continental assemblies, especially with the breakup of Pangaea at the end of the Permian. They also are coincident with extinction time lines, such as the major Permian extinction.

4.2. Structural control of Zechstein saline emplacement

A deep-seated origin for Zechstein salines implies that there would be feeder structures that contain salt dikes and other features that demonstrate upward flow from underlying deep sources. The feeder architecture is similar to that proposed by Blundell and others [58], who modeled a deep-seated brine influx into the Zechstein basin.

An example demonstrating the possible, exotic, deep source of the chemical, mud-brine volcanism was observed by one of the authors on a field trip to the Thomas Munzer shaft of the Sangerhausen Mine in Germany in May 1991. **Figure 18** shows steeply inclined and folded, anhydrite laminae superimposed across a relatively flat-lying halite (gray) – sylvite (pink-red)

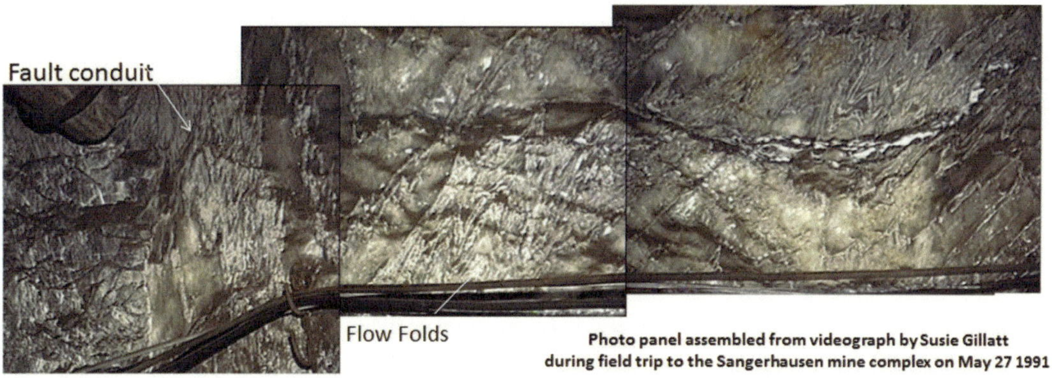

Figure 18. Flow-folded, anhydrite and salt layers exposed in a tunnel face at the Thomas Munzer shaft, Sangerhausen Mine, Germany. The left panel shows a vertical fault with the salt-anhydrite on the right over the Kupferschiefer strata. (Screen shots were assembled by Keith from a video by Susan Gillatt, taken May 27, 1991, on a field trip led by John Guilbert.).

saline deposit that had not experienced the isoclinal folding. In the middle panel, the anhydrite laminae cut a pink sylvitic layer. In the right panel, the pink sylvite layer appears to increasingly cross cut the anhydrite laminae and is intercalated with white anhydrite or gypsum. The folding cannot be tectonic, as both units should be equally affected by the fold deformation.

The folds and mutual cross-cutting relationships of **Figure 18** are consistent with soft, hydrothermal flowage of immiscible, saline brines from an underlying, density-zoned, saline, brine chamber. In this chamber, halite (NaCl specific gravity of 2.16) – sylvite (KCl specific gravity of 1.99) brines are unmixed from anhydrite ($CaSO_4$ specific gravity of 2.97) brine in a density-stratified brine chamber. The lower density halite-sylvite brine is expelled first and is quickly followed by expulsion of the higher density anhydrite brine, which is injected as cross-cutting laminae into the halite-sylvite brine while it is still a semi-liquid, partially solidified, colloidal gel. Then, both are deformed by flow flowage away from the fault feeder at the left side of **Figure 18**. Both the early halite and the later, isoclinally folded, anhydrite formations display vertical foliation and axial planes within an apparent, fault-related conduit that has a sharp contact on its left side with the Kupferschiefer black shale section, as shown in the left panel of the photo.

Kinematics of the broad folding in the halite and the asymmetric isoclinal folding in the anhydrite in **Figure 18** are consistent with outward flowage to the right into progressively more horizontal flow layering from the fault feeder (conduit) at the left side of the figure. The overall occurrence of the saline brine is consistent with it being a late fractionate of deep-seated, higher density, mud-brine plume that created the high temperature, Kupferschiefer carbonaceous muds. There are exploration possibilities related to the association of fault-controlled brine plumes with veins containing high-grade copper-silver-(PGE-Au)-carbonaceous material.

4.3. Zechstein dolomites as a chemical carbonate mud

Zechstein dolomites occur beneath anhydrite units throughout at least a threefold chemical, sedimentary, cyclothem-like sequence. In the lowest or earliest sequence, the dolomites and calcitic dolomites transition downward into the carbonaceous shales of the Kupferschiefer.

This downward chemical stratigraphy culminates in a very kerogen-rich, copper-rich, illite-rich, shale at its base. The Kupferschiefer itself interfingers laterally with a unit widely referred to as the Boundary Dolomite [9], which rests on the top of Weissliegend silica mounds (see Figure 4 in [10]). The Boundary Dolomite is a dolomite-cemented sandstone that rapidly grades downward into the main, structureless sandstone of the Weissliegend. Thus, there appears to be a direct chemical and sedimentological connection between the Weissliegend chemical extrudites and the Boundary Dolomite, which is a magnesium carbonate fractionate at the top of the Weissliegend. In summary, the Zechstein dolomites can be portrayed as chemical sedimentary muds that precipitated above the minimum, dolomite-precipitation threshold of 40°C [59] in the upper portions of fault-controlled, mud conduits.

At the sites where the deep-sourced, mud-brine plumes emerge onto the sea bottom into seawater, sedimentary processes of settling operate. However, the brines themselves are the result of a much deeper chemical process that is not evaporative in nature. This deeper origination process also pertains to the extensive dolomites that typically form the base of a given Zechstein cycle between the anhydrite units and the overlying, sodium-rich, saline deposits. Potassium-rich saline minerals (sylvite) can appear at the end of a given cycle. The chemical fractionation sequence is dolomite, then anhydrite, then halite, and finally sylvite. The exotic magnesium-rich minerals (such as talc and serpentine) typically occur in the early, dolomite to anhydrite part of the sequence, as pointed out by Warren [57].

The above mineral sequence cannot be formed by precipitation through evaporation of normal seawater. The normal precipitation sequence for precipitation from seawater is aragonite, then gypsum, then halite, and then very small amounts of bittern as magnesium and potassium chlorides. Dolomite is conspicuous by its absence. The precipitation or reaction sequence from a deep-sourced, hydrothermal, Mg-K-Cl brine is calcite, then dolomite, then anhydrite, then halite, and finally large amounts of sylvite and MgCl (for example tachyhydrite [$CaMg_2Cl_6 \cdot 12H_2O$] and bischofite [$MgCl_2 \cdot 6H_2O$]). In the UDH perspective, the Mg-K-Cl brine component is a low-density, hypersaline component that separates near the end of a density-driven, brine-mud, fractionation sequence. This Mg-K-Cl brine component would reach the undersea surface and be deposited by settling out of a brine plume.

5. Rote Fäule

The hematite-enriched facies (the Rote Fäule) (**Figure 19**) is wedge-shaped, traverses the stratigraphic sequences of the Zechstein base, and produces a lateral zoning of different ore types (**Figure 5** and the block diagram of Figure 101 in [11]). As a result of the acute-angled dip, economic ore-concentration occurs at the border between Rote Fäule and the underlying sandstones and continues into the Kupferschiefer and Zechstein. The hematitic mineralization is the early stage of Rote Fäule and is described as diagenetic oxidization stage (DOS) by Piestrzyński and Sawlowicz [13]. The DOS is cross cut by a later stage of noble element metallization (Au-Pt-Pd)-U referred to as secondary oxidization stage (SOS) [13].

From the perspective of the UDH model, both DOS and SOS stages are viewed as evolving, strongly oxidized, low-sulfur, chlorine-rich, hydrothermal plume that is superimposed on

Figure 19. The Rote Fäule, with typical spotted hematite texture overprinting the basal Zechstein dolomite (from Hitzmann [25]).

pre-existing, reduced, Kupferschiefer-Zechstein metallization. In this context, the noble element metallization is viewed as a late, fractionated product that typically forms near the edge of the Rote Fäule plume. The emplacement of the oxidized Rote Fäule plume was independent of the reduced, pre-existing Kupferschiefer Cu-Ag mineralization. Each event represents a separate pulse of an evolving, deep-seated, mud-brine plume system. Each pulse had its own internal evolution that was independent of the other pulses. However, they are broadly connected in space and time at the scale of the entire crust, in the context of a rift system that taps ultramafic sources in the mantle.

In the Polish Kupferschiefer-Zechstein, hematitic and silicic alteration in the footwall sandstone units appears to be spatially associated with the Odra fault system shown in **Figure 15**. This fault system is inferred to have tapped deep-seated, metal and kerogen sources in the underlying basement. As with the Kupferschiefer, the highly anomalous nature of the combined Cu-Ag-Pb-Zn-Mo-Au-PGE-Ni-V-Cr geochemistry strongly suggests a deep-seated, UDH, serpentinized source in the basement from which mud-sulfide brines moved to the seafloor interface via a network of deeply penetrating basement cracks. Unlike the Kupferschiefer, the main mineral in the Rote Fäule is hematite, which indicates a much more oxidized oxidation state compared to the reduced Kupferschiefer. A more oxidized character of the Rote Fäule is in accord with the more oxidized nature of the Zechstein carbon-oxygen-rich carbonates.

The Rote Fäule hematitic zone typically occurs in the footwall of the Kupferschiefer, but locally overprints the reduced facies, including the hanging wall Zechstein dolomite marls above the Kupferschiefer shale. The Rote Fäule hematitic overprint shown in **Figure 19** suggests the main Rote Fäule hydrothermal plume was oxidized and its continued 'pumping' resulted in an 'overprint' of the earlier, reduced facies. In some cases, lithic clasts of reduced Kupferschiefer ore appear as included clasts in the Rote Fäule in the northwest part of the Lubin deposit as shown in Figure 3 in [60] and Figure 3b in [61]. These figures show that the reduced Kupferschiefer xenoliths are earlier than and are being consumed by hematitic Rote Fäule mineralization.

An important additional observation is that the gold is spatially associated with Rote Fäule and also cross cuts and replaces earlier, reduced copper sulfides (such as bornite, chalcocite, digenite, and covellite) that are hosted in the kerogenous Kupferschiefer black shale, as shown in the above cited figures in [60, 61]. The gold is also strongly correlated with other noble elements, such as platinum and palladium. Locally, uranium, as thucholitic kerogen, is also tied to the noble element association (Figures 2, 5, 6 and 9 in [18]). The noble element-uranium association is best developed in what is widely referred to in the above literature as the transition zone between the oxidized Rote Fäule and the reduced Kupferschiefer black shale. From a UDH perspective, the gold-noble element-uranium-Rote Fäule-related mineralization is inferred to be a younger overprint that destroys pre-existing, reduced Kupferschiefer mineralization, and possibly remobilizes it to the margins of the Rote Fäule bodies, where it is co-spatial with the noble element-uranium metallization. As such, the noble element-uranium association is not considered to be directly genetically related to the Kupferschiefer copper-silver metallization, which is earlier.

The previous point has exploration implications in that the margins of Rote Fäule bodies may be prospectable for independent, noble element-uranium accumulations. Potassium-argon dating of high to moderate quality illites, extracted from the Rote Fäule [42], yield four samples that are considered viable indicators of the age of the Rote Fäule. As developed above, the average age of the Rote Fäule as recorded by the more reliable illite ages is 244.5 $\pm$ 3.3 Ma (**Figure 17**). These ages are consistent with the observation that the Rote Fäule cross cuts the lower Zechstein Werra anhydrite unit as depicted in **Figure 5**. As such, the Rote Fäule and its attendant noble element-uranium mineralization is inferred to be a late stage, oxidative feature of the Zechstein-Kupferschiefer hydrothermal diapir that is coincident with the early expulsion of salines.

From the perspective of the UDH mud-volcanic model, the Rote Fäule represents an independent hydrothermal event that is not directly related to the earlier Kupferschiefer-Zechstein deposits. Wherever it is observed, the Rote Fäule cross cuts the earlier Kupferschiefer-Zechstein deposits and, so far as we are currently aware, does not exhibit chemical, extrudite facies. It is possible that the Rote Fäule may merge or fractionate into one of the upper Zechstein saline cycles. Nevertheless, its consistent position in the paragenetic sequence and its consistent spatial correspondence with the Kupferschiefer-Zechstein metallization suggests that the Rote Fäule is a late hydrothermal pulse of the broader chemical mud-volcanic process, which ranges in age from about 265 to 235 Ma. More to the point, the Rote Fäule contains its own metallogenic signature that is derived from hydrothermal fractionations within the Rote Fäule event and is not dependent on remobilization of metal sources in earlier formations or pre-existing Kupferschiefer and Zechstein metal deposits.

The three pulses in the broader Kupferschiefer-Zechstein metallization sequence are represented by (1) Weissliegend-Kupferschiefer Cu-Ag (Re, Pb) metallization at ~265–255 Ma, (2) Zechstein Zn-Cu-Pb-Ag metallization at ~250–245 Ma, and (3) Rote Fäule Au-(PGE-U-Co-Se) metallization at ~245–235 Ma. The three heat pulses may represent different stages of dehydration in the underlying ultramafic basement.

6. Later (Rücken) veins

Persistent attempts to determine the age of the Kupferschiefer have resulted in the identification of numerous overprints that are younger than the main Kupferschiefer event that occurred

http://dx.doi.org/10.5772/intechopen.72560

Figure 20. Late siderite-barite-Co-Ni-(Cu, U) veins associated with late faults displacing the Kupferschiefer.

between 265 and 240 Ma. Periodically, the system is re-energized by pulses of new, mantle-sourced, heat flux. One of the more well-known events is the Late 'Alpine age', polymetallic 'Rücken' veins that cross cut the Kupferschiefer-Zechstein metal deposits. **Figure 20** is a photograph of siderite-barite-Co-Ni (Cu, U) vein associated with late faults that displace the bedding of the Kupferschiefer. 'Rücken' veins are especially common in the Mansfeld-Sangerhausen Districts, where paleomagnetic ages on Rote Fäule hematite have yielded a circa 150 Ma paleomagnetic age. This age is inferred to represent heating and resetting of an early Triassic Rote Fäule event by latest Jurassic age, regional heating associated with the 'Rücken' vein event.

The region remains the site of high heat flow, as shown by the geothermal field of the southern Lausitz basement in the Spremberg region (**Figure 21**) and a well-defined, heat flow anomaly and helium anomaly east of Lubin (Figure 4 of [58]). In the tectonic setting of crustal rifting, deep lineaments opened that are part of a global-scale fracturing event whose traces remain to the present. This fracturing opened pathways for brines from very deep-seated mantle locations. Hot, hydrothermal fluids ascended upwards and expelled material with a basic to ultrabasic chemical signature along the fractured, rifting lineaments.

7. Geotectonic setting

The stage for the Zechstein-Kupferschiefer event was set by the assembly of Pangaea, which was mostly complete by the end of the Carboniferous. With this blockage of heat flow from the

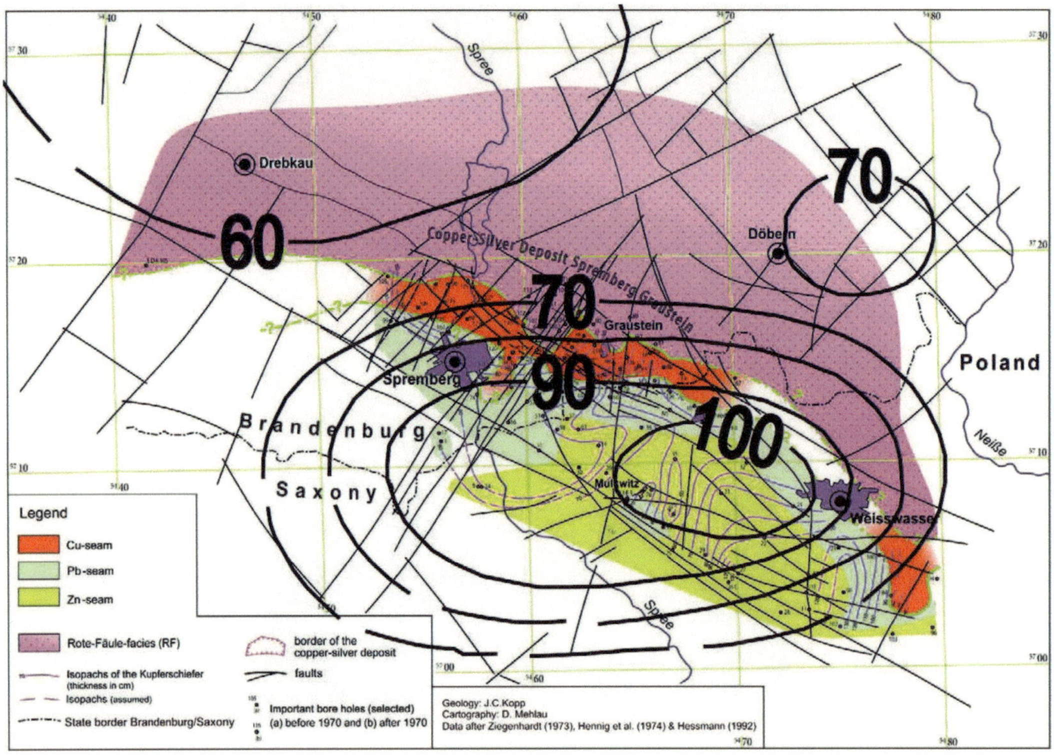

Figure 21. Geothermal field of the southern Lausitz basement showing temperature isoclines and metal zones (from Kopp et al. [27]).

mantle by a large continental mass, hot spots began to develop to destabilize the continental crust and create rifts. The creation of the Zechstein-Kupferschiefer hot spot at the end of Permian time created what amounts to a long-lived zone of weakness that has been periodically active since the main event that formed the Zechstein-Kupferschiefer system. This zone is currently evident as a thermal anomaly centered on the Spremberg area of Germany (**Figure 21**). The presence of the current geothermal anomaly indicates the compliance with a requirement for a massive amount of energy that is needed to drive processes like the one that created the Zechstein-Kupferschiefer system. Smaller scale energy models, such as convection cells, have a problem scavenging the energy required to leach a given source, let alone finding the fracture framework with which to access the leach candidate and remove its chemical contents and then migrate them to a trap site. The deep-sourced, UDH process accesses an endless amount of mantle-derived, conductive heat flow that migrates straight upward along a simple, structurally determined, conduit system.

An elongated intra-continental basin began subsiding in the Late Permian, as marked by the deposition of the Upper Rotliegend at approximately 265 Ma. At this time, the Upper Rotliegend distribution was very similar to the Zechstein-Kupferschiefer distribution (Figure 1 in [5]) and is inferred to be related to the same aulacogenic process. During the rift phase from the latest Early Permian until the mid-Triassic, Zechstein sediments and salines were deposited. Basement fractures are good places for vertically focusing large amounts of deep-sourced materials through basement-rooted pipes and chimneys. The source of the base metal deposits can be explained by deep-penetrating faults, which opened channel ways for ascending fluids. These structures were active during different stages of the post-Variscan basin evolution. Subsequent phases of localized subsidence continued from the mid-Triassic to Jurassic period of extension [27].

The high-grade copper mineralization of the Zechstein base occurs exclusively in the area of the stratigraphic border between the Saxo-Thuringian and Rheno-Hercynian shear zones of the Variscan orogeny. The so-called European Copper Belt, with its high-grade gold and polymetallic mineralization on the southern rim of the Kupferschiefer Sea (Figure 4 in [62]), may have been fed through a pre-existing suture zone of the Central European Variscides, which is bordered on its southern flank by deep-reaching faults, such as the Odra fault system [27, 63]. These faults form the southern boundary of the Zechstein-Kupferschiefer basin.

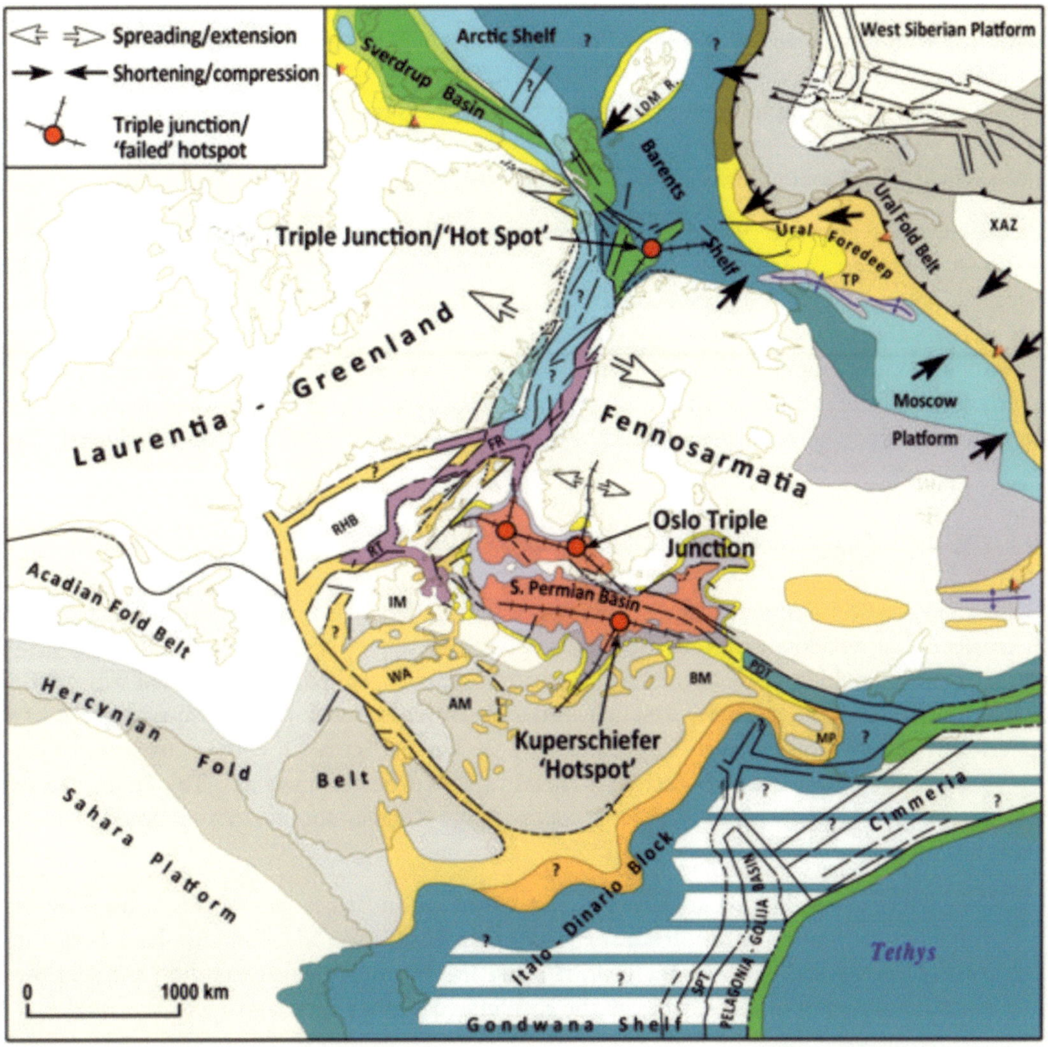

Figure 22. Location of the triple junction/ 'hot spot' near the Permo-Triassic horizon (255–245 Ma) (modified from Ziegler [65]).

Subsidence within the Zechstein-Kupferschiefer basin (which includes the upper Rotliegend) may have been aided by withdrawal of massive amounts of brine-mud slurries from beneath the current centre of the basin. It is no coincidence that thickness anomalies in the Zechstein coincide with subsidence anomalies on the Zechstein-Kupferschiefer paleosurface (Figure 2 in [64], and Figure 2 in [27]).

The plate tectonic configuration in the Kupferschiefer area at the Permo-Triassic boundary (circa 255–245 Ma) consisted of a triple junction 'hot spot' (**Figure 22**). At that time, Laurentia – Greenland was moving west-northwestward away from Fennosarmatia-Baltica [65] in a rift setting. Also at that time, the Fennosarmatia-Baltica and the Moscow Platform were moving northeastward and colliding with the Ural Fold belt. At the time of the Permian-Triassic boundary, the area of the Zechstein-Kupferschiefer was at a triple junction between a mid-ocean rift and the Ural fold belt. The Zechstein basin was an area of crustal extension. Basement mud-brine slurries migrated upward along subvertical fault zones and were periodically emplaced as the rock series of some of the Upper Rotliegend, Kupferschiefer, and Zechstein carbonates. Not all of the rocks in the Zechstein-Kupferschiefer system are considered to have a deep origin. For example, the red hematitic Rotliegend sandstones described by Glennie [6] are considered to be oxidized eolian sandstones deposited in a desert environment and have undergone normal sedimentary diagenetic processes.

The paleostress regime inferred for the greater Zechstein-Rotliegend region in Permo-Triassic time is shown in **Figure 22**. The regional stress field inferred from **Figure 22** was one of NNE-SSW compression and WNW-ESE extension. In this far-field stress regime, features such as the Oslo graben opened in extension, whereas a number of the WNW features that control reef trends and oil and gas accumulations may have operated as reverse faults. The more northwestern features, which are common features in the Polish part of the Zechstein basins, may have moved in right transpressive slip, whereas less continuous, NE-striking connectors may have operated in left slip.

8. Origin of Zechstein-Kupferschiefer system

The entire Zechstein-Kupferschiefer system can be explained as a product of ultra-deep hydrothermalism that developed during the process of steatization of serpentinite in the lower crust above or near possible mantle hot spots. One such hot spot is posited to have existed beneath the Rote Fäule anomaly in the Berlin area. An interpretive cross-section of the emplacement of a generic ultra-deep hydrothermal (UDH) system is shown on **Figure 23**, with the uppermost part representing the Kupferschiefer-Zechstein metal and hydrocarbon system. The UDH process that created the Kupferschiefer-Zechstein copper-silver deposit supersystem proceeded in six major steps.

1. The UDH process begins at a zone of depressurization that is tectonically caused by the stress regime shown in **Figure 22**. The depressurization induces mantle heat flow into zones of weakness, such as the Oslo rift and the north-northeast-trending rift arm that extends south-southwestward from the Kupferschiefer hot spot.
2. Mantle heat flow dehydrates the Serpentosphere [66], at the Mohorovicic discontinuity, where it had accreted at the mantle-crust boundary during numerous, previous, flat

subduction events that range back in time into the Proterozoic. The influx of mantle heat induced dehydration of serpentinite into talc-carbonate rock (steatite). Mass balance of this reaction indicates massive amounts of excess water are created, which combines with other chemicals (such as Na, Cl, K, etc.) to produce high-density mud-brines that are enriched in trace metals typical of serpentine, such as PGE, Cr, Ni, and Co. The stratigraphy of the combined Weissliegend-Kupferschiefer-Zechstein indicates that three major heat/dehydration pulses might have operated.

3. The chemical and mineralogical reactions in the UDH process take place at high temperatures and pressures that attend the generation of supercritical, non-ionic fluid with a significant component of nano- to micro-sized particulates in the lower crust in the vicinity of the steatite reaction.

4. The brines are expelled due to extremely high fluid overpressures that developed during steatization. This pressure release, along with the exothermic heat generated during the serpentinization reaction, causes upward migration of the buoyant brines toward low pressure anomalies associated with deep-seated basement faults. The dense brine slurries then rise into higher levels in the basement, where they cross the subcritical boundary for brine water at around 400°C and ionize. At this point, mineral compounds start to form from less molecular, flocculate-like particulates. For example, sulfide melt droplets and emulsates may start crystallizing into fibrous, spherulitic minerals (such as shown in **Figure 7**).

5. The high-density brines then emerge onto the paleosurface (inland seabed or paleotopographic surface), forming extensive, saline playa pans and pancake-like mud volcanoes near where the conduits intersect the paleosurface. Basin subsidence may be aided by the withdrawal of massive amounts of mud-brine from chambers deep in the crust near the steatization reaction chambers.

6. The brines included a primary kerogen component that was inherited from the serpentinite source. This kerogen was progressively hydrogenated by the hydrothermal process until it entered the oil generation window near the Permian-Triassic boundary circa 252 Ma. A considerable amount of petroleum may have accompanied the emplacement of the greater Kupferschiefer-Zechstein mineralization. Hydrothermal oil generation accompanied precipitation or deposition of metal sulfides from H_2S-rich brines, which released large amounts of hydrogen. The hydrogen immediately hydrogenates the adjacent kerogen hydrocarbon, especially at illite-phengitic muscovite interfaces. Continued hydrogenation under such a process led to hydrothermal oil generation.

The above-described emplacement process is associated with several associated process-related phenomena.

The rising mud-brine plume fractionates along reaction paths governed by density, composition, decreasing temperature, and decreasing pressure. The density and composition determine the order in which the products are created in a given reaction. The result is a systematic, paragenetic sequence with the sequential, but overlapping, formation of by-product non-metallic minerals (dolomite, then quartz and illite and oil, then later dolomite, then calcite). The sequential expulsion of these by-product minerals is vertically and horizontally zoned. The metallic mineral

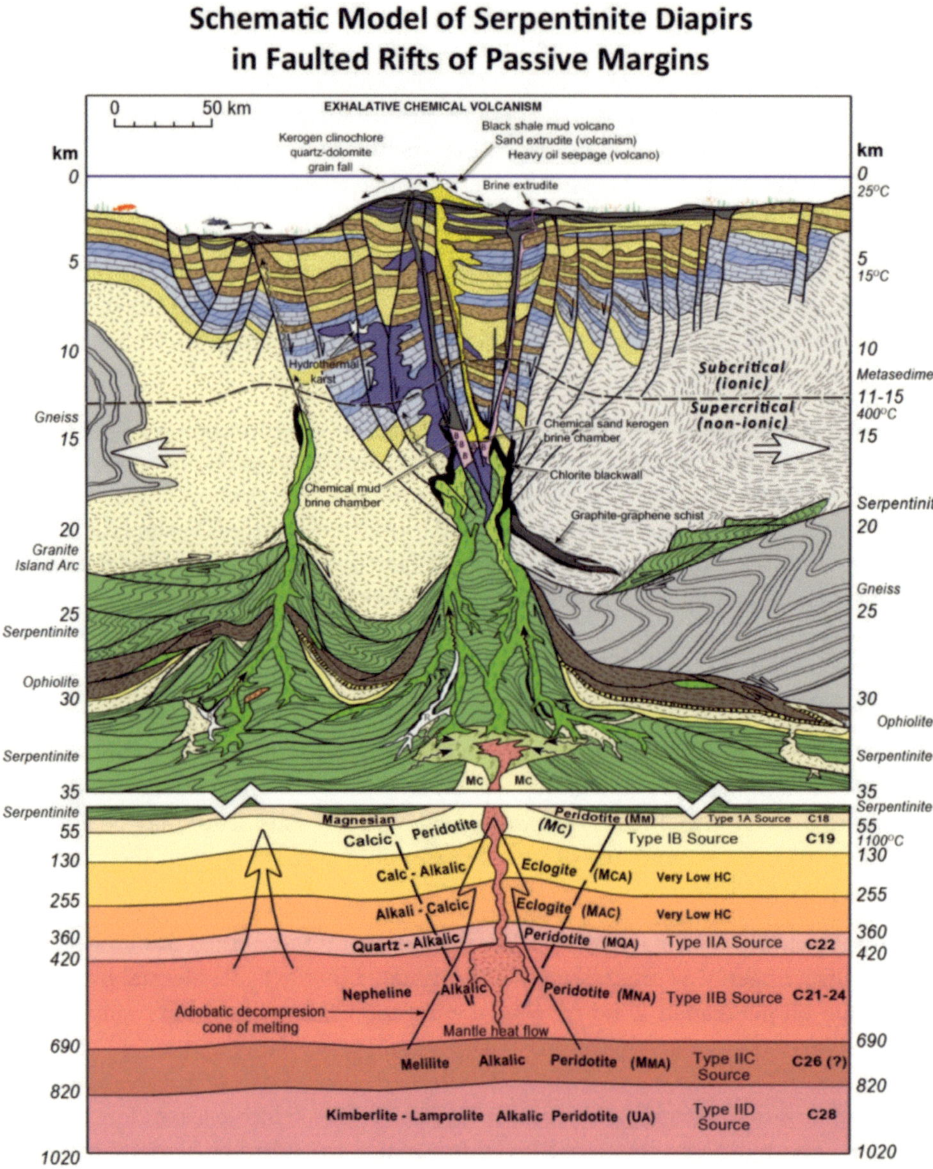

Figure 23. Schematic model of serpentinite diapirs in faulted rifts of passive margins. The model is generic and does not apply to the Zechstein-Kupferschiefer stratigraphy in detail. The part that pertains to the Zechstein-Kupferschiefer system is outlined in red and applies to deposition of the main Kupferschiefer just prior to the Zechstein chemical mud-volcanic event that formed the anhydrite-halite-potash lithocap over the Kupferschiefer.

sequence is generally high digenite, high chalcocite, bornite, covellite, orange bornite, yarrowite, low chalcocite, low digenite, djurleite, and anilite. Temperature ranges of these sequences are from 600°C to less than 40°C. Pyrite can form variously through the sequence, but is especially well developed late in the sequence. Sulfur isotopes generally indicate that the sulfide sequence becomes progressively oxidized during its ongoing pulsed emplacement over the 30 million year interval.

The emplacement of the Zechstein-Kupferschiefer chemical system produced regional scale, proximal to distal, compositional patterns (**Figure 15**). The resulting pattern left systematic mineral and petroleum signatures in the strata surrounding the fractures that can be used to vector toward a probably deep-sourced, mud vent that comprises the central energy and metal anomaly.

The massive amounts of brines and acids would have been toxic to any life immediately above the eruption site and would have entombed any swimming animals above it, such as the pickled, metallized herring fossil in **Figure 24**. The gigantic mud-brine plumes created a massive chemical imbalance in the Earth's ecosystems that induced the well-known Permian extinction event. The toxicity would have been maximized near the tops of the reductive black shales. The black shales mark the extinction event below the Zechstein at the top of the Kupferschiefer black shale (**Figure 3**). The toxicity would have continued into the saline intervals that mark the upper part of the Zechstein sequence. However, the global ecosystem rapidly recovered in the Triassic immediately after the reduction event and salinity crisis ended.

Hence, the Kupferschiefer and its larger Zechstein-Rotliegend system provide clues, not only into metallogenic processes, but also into the entire geologic process that operated at the end of Permian time during the breakup of Pangaea and the great extinction event.

Figure 24. Fish fossil (*Palaeniscum Freiesiebeni*) in the Kupferschiefer replaced by chalcopyrite and chalcocite, Lubin, Poland. Photo by Volker Spieth.

9. Summary

From a UDH, chemical mud volcanism point of view, the following major points are critical to the reconsideration of Zechstein-Kupferschiefer mineralization:

1. The black shale component of the Kupferschiefer is a small part of the co-genetic, mud-chemical, volcanic system that began with the Weissliegend silica extrudite complex circa 265 Ma and ended with the Rote Fäule oxidative event circa 235 Ma. As such, the Weissliegend to Kupferschiefer-Zechstein sequence was long-lived and represents a phased/pulsed, high-energy to low-energy continuum that is punctuated by quiescent periods.

2. The deep-seated nature of the generic Zechstein-Kupferschiefer-Weissliegend system is manifested by exotic element chemistry (PGE, Co, Ni, Cr, and V) that indicates a deep, ultimately serpentinized, peridotite source in the lower crustal Serpentosphere (Moho).

3. The exotic nature of the chemical constituents allows a much more flexible explanation of genetic issues, such as disparate age dates. In the case of the age dates, a given sulfide

mineral might equilibrate at different depths and temperatures with different isotopic signatures that would lock in different ages. Similarly, a given mineral, such as high digenite, might crystallize at 600°C deep in the vent complex, whereas a nearby mineral, such as anilite, would stabilize at very shallow levels in the complex at below 40°C. Phengitic, highly crystalline pure muscovite would lock in its K-Ar clock at 350°C, whereas cooler, more smectitic, illitic muscovite would be much less Ar-retentive and indicate temperatures below 200°C.

4. The Zechstein-Kupferschiefer, hydrothermal, chemical mud-brine, mud-volcanic system represents a continuum process that operates from high-energy to low-energy. The various exotic minerals within this system capture different parts of the process. As such, the mud-brine volcanic system requires modeling as an open, dynamic, non-equilibrium system, not a closed, static, equilibrium system, as is tacitly assumed by many other models.

5. Most of the Cu-Ag resource is now established to be hosted in the Weissliegend unit and not in the traditional Kupferschiefer black shales. As such, exploration should focus on silica-mud mounds that may represent energy centers for the Weissliegend, silica-Cu-Ag-related, mud volcanism.

6. Reconsideration of the textural features of the Kupferschiefer shale indicate that the entire unit contains exotic clasts and matrix and that the unit includes a high temperature stage that formed at least 10 km beneath the paleosurface. The Kupferschiefer-Zechstein mud slurries flowed onto the paleosurface into low-relief depressions adjacent to Weissliegend, silica extrudite, mud volcanoes. The Kupferschiefer is also a dense, reduced, carbonaceous slurry that separated from a dolomite-cemented, silica extrudite (the Boundary Dolomite) that formed on the top of the silica mounds (especially in the Lubin district of southwest Poland).

7. The presence of exotic minerals (such as talc, serpentine, and clinochlore) in the overlying Zechstein salines, indicate these minerals were derived from a deep, steatized, serpentinite source beneath the Zechstein basin.

8. Zechstein basin subsidence may have been aided by withdrawal of massive amounts of chemical, high-density brines from steatitic reaction chambers in the lower crust.

9. The kerogen in the Kupferschiefer black shales is also considered exotic and was hydrogenated and augmented during its upward ascent and was reacted into hydrogen-rich alkane oils and low C-number alkane gases (especially methane).

10. Based on reconsideration of available illite K-Ar data and Re-Os isotope data, the main Kupferschiefer-Zechstein mineralization is inferred to have occurred between 265 Ma and 235 Ma. The Permian-Triassic boundary is located at the top of the Kupferschiefer shale at its rapid transition to Zechstein carbonates at circa 252 Ma.

11. The oxidized, hematitic, Rote Fäule unit represents a younger hydrothermal plume that was superimposed on all of the earlier chemical exhalatives and that has yet to have a recognized, mud-volcanic component. Gold (PGE, Co, U, and Se) characterize the metallogeny of this later event and were deposited as a secondary, oxidized fractionate (SOS) that is indigenous to the Rote Fäule.

12. The geotectonic setting for the Kupferschiefer was emplacement into an aulacogen, failed rift along deep-seated, west-northwest-trending faults, for example the Odra fault system. The overall shape of the Zechstein basin suggests a Kupferschiefer hot spot was associated with a major anomaly in the Rote Fäule in the vicinity of Berlin.

13. Toxicity related to the emplacement of the reduced black shale of the Kupferschiefer was focused at the Permo-Triassic boundary. The toxic, brine plumes of the Kupferschiefer contributed to the great Permian extinction event that attended the beginnings of the breakup of the Pangaea supercontinent.

The Zechstein-Kupferschiefer super system is considered to be an archetypal example of regional scale, mud volcanism related to hydrothermalism that was sourced in serpentinized peridotite and that was triggered by mantle heat flow. Other examples of known or possible, regional scale, mud volcanism include the Permian basin of northern Texas, USA; the Green River basins of Eocene age in central Colorado, USA; the current Mariana serpentine mud-volcanic system in the 700 km long, 60 km wide forearc of the Marianas; and the current hydrocarbon-related, mud volcanism in the Caspian Sea region and the Salton Sea in southern California, USA.

The Kupferschiefer-Zechstein and its analogues point to the importance of an Earth-scale process of hydrothermalism that has largely been underestimated in the geologic literature. In its own right, hydrothermal, chemical mud volcanism may be as important a geologic process as metamorphism, magmatism, and sedimentation.

Author details

Stanley B. Keith[1], Volker Spieth[2,3] and Jan C. Rasmussen[4]*

*Address all correspondence to: geoarizona@gmail.com

1 MagmaChem Exploration, Sonoita, Arizona, United States

2 VS.GLOBALMETAL LLC, Tucson, Arizona, United States

3 University Stuttgart, Germany

4 Jan Rasmussen Consulting & University of Arizona, United States

References

[1] Kucha J. Precious metal bearing shale from Zechstein copper deposits, Lower Silesia, Poland. Transactions Institution Mining and Metallurgy, Section B Applied Earth Science. 1982;**92**:B72-B79

[2] Zientek ML, Oszczepalski S, Parks HL, Bliss JD, Gorg G, Box SE, Denning PD, Hayes TS, Spieth V, Taylor CD. Assessment of undiscovered copper resources associated with the Permian Kupferschiefer, Southern Permian Basin, Europe. U.S. Geological Survey Scientific Investigations Report. 2015; SCI-2010-5090-U. pp. 1-94

[3] Gautier DL. Carboniferous-Rotliegend total petroleum system description and assessment, results summary. U.S. Geological Survey Bulletin. 2003;**2211**:1-28

[4] Alderton DHM, Selby D, Kucha H, Blundell DJ. A multistage origin for Kupferschiefer mineralization. Ore Geology Reviews. 2016;**79**:535-543

[5] Glennie KW, Buller AT. The Permian Weissliegend of NW Europe: The partial deformation of Aeolian dune sands caused by the Zechstein transgression. Sedimentary Geology. 1983;**35**:43-81

[6] Glennie KW. Permian Rotliegendes of northwest Europe interpreted in light of modern desert sedimentation studies. American Association of Petroleum Geologists Bulletin. 1972;**56**:2048-2071

[7] Hurst A, Cartwright JA, editors. Implications for Hydrocarbon Exploration and Production. American Association of Petroleum Geologists Memoir; 2007. pp. 1-274

[8] Blaszczyk JK. Paleomorphology of Weissliegendes top as control on facies variability in ore-bearing series of Lubin copperfield (in Polish). Geologica Sudetica. 1981;**16**:196-214

[9] Krasoñ J. Podzia stratygraficzny cechsztynu pónocnosudeckiego w oewietle badañ facjalnych. (In Polish). Geologia Sudetica. 1964;**1**:221-255

[10] Piestrzyński A, Sawlowicz Z. Esploration for Au and PGE in the Polish Zechstein copper deposits (Kupferschiefer). Journal of Geochemical Exploration. 1999;**66**:17-25

[11] Pieczonka J, Piestrzyński A, Mucha J, Gluszek A, Kotarba M, Więclaw D. The red-bed-type precious metal deposit in the Sieroszowice-Polkowice copper mining district, SW Poland. Annales Societatis Geologorum Poloniae. 2008;**78**:151-280

[12] Oszczepalski S. Origin of the Kupferschiefer polymetallic mineralization in Poland. Mineralium Deposita. 1999;**34**:599-613

[13] Piestrzyñski A, Pieczonka J, Gluszek A. Redbed-type gold mineralisation, Kupferschiefer, South-West Poland. Mineralium Deposita. 2002;**37**:512-528

[14] Radliński M, Sawłowicz Z. Altered rhythmic sulphide bands in the Weissliegend sandstone in the Rudna Mine, Fore-Sudetic Monocline, Poland. Biuletyn Państwowego Instytutu Geologicznego. 2017;**468**:61-78

[15] Sawlowicz Z, Wedepohl KH. The origin of rhythmic sulphide bands from the Permian sandstones (Weissliegendes) in the footwall of the Fore-Sudetic "Kupferschiefer" (Poland). Mineralium Deposita. 1992;**27**:242-248

[16] Alt JC, Schwarzenbach EM, Früh-Green GL, Shanks III WC, Bernasconi SM, Garrido CJ, Crispini L, Gaggero L, Padrón-Navarta JA, Marchesi C. The role of serpentinites in cycling of carbon and sulfur: Seafloor serpentinization and subduction metamorphism. Lithos. 2013;**178**:40-54

[17] Lewan MD, Kotarba MJ, Wieclaw D, Piestrzynski A. Evaluating transition-metal catalysis in gas generation from the Permian Kupferschiefer by hydrous pyrolysis. Geochimica et Cosmochimica Acta. 2008;**72**:4069-4093

[18] Kucha H, Przbylowicz W. Noble metals in organic matter and clay-organic matrices, Kupferschiefer, Poland. Economic Geology. 1999;**94**:1137-1162

[19] Pašava J, Oszczepalski S, Du A. Re-Os age of non-mineralized black shale from the Kupferschiefer, Poland, and implications for metal enrichment. Mineralium Deposita. 2010;**45**:189-199

[20] Michalik M. Chlorine containing illites, copper chlorides and other chlorine bearing minerals in the Fore-Sudetic copper deposit (Poland). In: Papunen H, editor. Mineral Deposits. Rotterdam: Balkema; 1997. pp. 543-546

[21] Large DL, MacQuaker J, Vaughan DJ, Sawlowicz Z, Gize AP. Evidence for low-temperature alteration of sulfides in the Kupferschiefer copper deposits of southwestern Poland. Economic Geology. 1995;**90**:2143-2158

[22] Püttmann W, Hageman HW, Merz C, Speczik S. Influence of organic material on mineralization processes in the Permian Kupferschiefer Formation, Poland. Advances in Organic Geochemistry Organic Geochemistry. 1988;**13**(1–3):357-363

[23] Kucha H. Feldspar, clay, organic and carbonate receptors of heavy metals in Zechstein deposits (Kupferschiefer-type), Poland. Transactions Institution Mining and Metallurgy, Section B Applied Earth Science. 1985;**94**:B133-B146

[24] Borg G, Piestrzyński A, Bachmann GH, Püttmann W, Walther S, Fiedler M. An overview of the European Kupferschiefer deposits. Society of Economic Geologists, Special Publication. 2012;**16**:455-496

[25] Hitzmann MW. Kupferschiefer (Central Europe) sedimentary-rock hosted, stratiform, copper deposits. SGA Workshop 2013, Kitwe, Zambia. 2014. pp. 1-59

[26] Krol P, Sawlowicz Z. Massive copper sulphide mineralization in dolomite from the Lubin Mine, (Fore-Sudetic Monocline). Biuletyn Panstwowego Instytutu Geologicznego. 2017;**468**:29-48

[27] Kopp JC, Spieth V, Bernhardt H-J. Precious metals and selenides mineralisation in the copper-silver deposit Spremberg-Graustein, Niederlausitz, SE-Germany. Zeitschrift Der Deutschen Gesellschaft Fuer Geowissenschaften. 2012;**163**(4):361-384

[28] Bernardini GP, Corsini F, Mazzetti G, Firenze GT. Investigations in the Cu-S-Se system. Chemie der Erde. 1976;**35**:50-62

[29] Roseboom EH. An investigation of the system Cu–S and some natural copper sulfides between 25 and 700°C. Economic Geology. 1966;(4):641-672

[30] Vaughan DJ. Sulfide mineralogy and geochemistry. Reviews in Mineralogy & Geochemistry, Chantilly, Virginia, USA (Mineralogical Society of America & Geochemical Society). 2006;**61**:1-714

[31] Skinner BJ. Stability of the tetragonal polymorph of Cu2S. Economic Geology. 1970;**65**:724-730

[32] Potter RWII. An electrochemical investigation of the system copper-sulfur. Economic Geology. 1977;**72**:1524-1542

[33] Morimoto N, Koto K. Phase relations of the Cu-S system at low temperatures: Stability of anilite. American Mineralogist. 1970;**55**:106-117

[34] Morimoto N, Kullerud G. Polymorphism in digenite. American Mineralogist. 1963;**48**:110-123

[35] Morimoto N, Koto K, Shimazaki Y. Anilite, Cu_7S_4, a new mineral. American Mineralogist. 1969;**55**:106-117

[36] Schröcke H. Die Entstehung der endogenen Erzlagerstatten. Berlin: de Gruyer; 1986. pp. 1-877

[37] Barton Jr BP. Solid solutions in the system Cu-Fe-S; part 1: The Cu-S and CuFe-S joins. Economic Geology. 1973;**68**:455-465

[38] Lewan MD. Experiments on the role of water in petroleum formation. Geochimica et Cosmochimica Acta. 1997;**61**(17):3691-3723

[39] Jung W, Knitzschke G. Kupferschiefer in the German Democratic Republic (GDR) with special reference to the Kupferschiefer deposits in the Southeastern Harz foreland. In: Wolf KH, editor. Handbook of stratabound and stratiform ore deposits, Part II, Cu, Zn, Pb and Ag deposits. Vol. 6. Elsevier Scientific Publishing Company; 1976. pp. 353-406

[40] Rentzsch J, Knitzschke G. Die Erzmineralparagenesen des Kupferschiefers und ihre regionale Verbreitung. Freiberger Forsch-H.C. Leipzig (Verl. Grundstoffind.) 1968;**231**:189-211

[41] Stedingk K, Rentzsch J. General map of the deep underground low-lying energy resources and raw materials in Saxony-Anhalt, Sheet II—Potential of the ore and spar sulfide minerals, in Digitale geowissenschaftliche Landesübersichtskarten von Sachsen-Anhalt im Maßstab 1:400,000 [Digital geoscience maps state of Saxony-Anhalt at a scale 1:400,000]. Landesamt für Geologie und Bergwesen Sachsen-Anhalt [State Agency for Geology and Mining Sachsen-Anhalt] Halle/Saale. 2003. Scale 1:400,000 [In German]. ISBN: 978-3-929951-34-9

[42] Bechtel A, Elliott WC, Wampler JM, Oszczepalski S. Clay mineralogy, crystallinity, and K-Ar ages of illites within the Polish Zechstein Basin: Implications for the age of Kupferschiefer mineralization. Economic Geology. 1999;**94**:261-272

[43] Mikulski SZ, Stein HJ. Re-Os isotopic age of the Cu-Ag sulphide ore and its mineralogical and geochemical characteristic from the Lubin-Polkowice mining area (SW Poland). Biuletyn – Panstwowego Instytutu Geologicznego. 2017;**468**:79-96

[44] Rentzsch J. The Kupferschiefer in comparison with the deposits of the Zambian Copperbelt. In Bartholome P, editor. Gisements stratiformes et provinces cuprifères. Liege: Société Géologique de Belgique; 1974. pp. 395-418

[45] Smith DG, Crosby A. The regional and stratigraphical context of Zechstein 3 and 4 potash deposits in the British sector of the southern North Sea and adjoining land areas. Economic Geology. 1979;**74**:397-408

[46] Stewart FH. The petrology of the evaporites of the Eskdale no. 2 boring, east Yorkshire; pt. 1 – The lower evaporite bed. Mineralogical Magazine. 1949;**28**:621-675

[47] Stewart FH. The petrology of the evaporites of the Eskdale No. 2 boring, East Yorkshire; pt. 3 – The upper evaporite bed. Mineralogical Magazine. 1951;**29**:557-572

[48] Fabricius J. Natural Na-K-Mg-Cl solutions and solid derivatives trapped in euhedral quartz from Danish Zechstein salt. Chemical Geology. 1987;**61**:95-112

[49] Fuchtbauer H, Goldschmidt H. Ein Zechsteinanhydrit-Profil mit Einlagerungen von Montmorilloni und einer abweichenden Serpentinvaritat. Heidelberger Beitrii.ge zur Mineralogie und Petrographie, Aus dem Geologischen Laboratorium der Gewerkschaft Elwerath. Erdolwerke Hannover, Bd. 1956;**5**:187-203

[50] Bauerle G, Bornemann O, Mauthe F, Michalzik D. Origin of stylolites in upper Permian Zechstein anhydrite (Gorleben Salt Dome, Germany). Journal of Sedimentary Research. 1999;**70**(3):726-737

[51] Pusch M, Hammer J, Kus J, Klosa D, Thiemeyer N, Mingerzahn G. Macro- and microscale distribution of hydrocarbons in the Staßfurt Hauptsalz Gorleben salt dome. Zeitschrift der Deutschen Gesellschaft fur Geowissenschaften. 2014;**165**(1):3-14

[52] Niemann H. Untersuchungen am Grauen Salzton der Grube "Königshall-Hindenburg", Reyershausen bei Göttingen. Contributions to Mineralogy and Petrology. 1960;**7**:137-165 (in German)

[53] Hryniv S, Peryt TM. Strontium distribution and celestite occurrence in Zechstein (upper Permian) anhydrites of West Poland. Chemie der Erde. 2010;**70**:137-157

[54] Wachowiak J, Pieczka A. Motukoreaite from the Kłodawa Salt Dome, Central Poland. Mineralogical Magazine. 2016;**80**(2):277-289

[55] Bodine Jr MW Notes on the clay mineralogy of the Retsof Salt Bed, International Salt Company Mine, Retsof, New York. U.S. Geological Survey, Open-File Report. 1983;**83–718**:1-13

[56] Fabricius J. Studies of fluid inclusions in halite and euhedral quartz crystals from salt domes in the Norwegian-Danish Basin. In: Sixth International Symposium on Salt, Toronto, Ontario, Canada. Vol. I. 1983. pp. 247-255. Available from: http://www.worldsaltsymposium.org/sixth-salt-symposium/

[57] Warren JK. Evaporites, brines and base metals: Fluids, flow and 'the evaporite that was'. Australian Journal of Earth Sciences. 1997;**44**:149-183

[58] Blundell DJ, Karnkowski PH, Alderton DHM, Oszczepalski S, Kucha H. Copper mineralization of the Polish Kupferschiefer: A proposed basement fault-fracture system of fluid flow. Economic Geology. 2003;**98**:1487-1495

[59] Deelman JC. Low temperature nucleation of magnesite and dolomite. Nues Jarbuch fur Mineralogie Monatshefte. 1999;**7**:289-302

[60] Piestrzynski A, Wodzicki A. Origin of the gold deposit in the Polkowice-West Mine, Lubin-Sieroszowice mining district, Poland. Mineralium Deposita. 2000;**35**:37-47

[61] Oszczepalski S, Chmielewski A, Speczik S, Krzeminski P. The northwest-trending extension of the Lubin-Sieroszowice Cu-Ag deposit. Mineral Resources to Discover – 14th SGA Biennial Meeting 2017. Vol. 3. 2017. pp. 1159-1162

[62] Spieth V. Copper-silver-gold deposits: Germany's world class potential in the 21st century. Erzmetall. 2016;**69**:38-49

[63] Rentzsch J, Franzke HJ. Regional tectonic control of the Kupferschiefer mineralisation in Central Europe. Zeitschrift fur Geologische Wissenschaften. Berlin (Verlag fur Geowissenschaften Berlin). 1997;**25**:121-139

[64] Jowett EC. Genesis of Kupferschiefer Cu-Ag deposits by convective flow of Rotliegende brines during Triassic rifting. Economic Geology. 1986;**81**:1823-1837

[65] Ziegler PA. Evolution of the Arctic-North Atlantic and the Western Tethys – A visual presentation of a series of paleotectonic maps. American Association of Petroleum Geologists (Memori). 1988;**43**:1-197

[66] Keith SB, Swan M, Rueslatten H, Johnsen HK, Page N. The serpentosphere. Geological Society of Nevada Newsletter. 2008;**23**(3):3

Chapter 3

Lead-, Zinc-, and Iron-Sulfide Mineralization from Northern Iraq

Ali Ismail Al-Juboury, Waleed S. Shingaly, Elias M. Elias and Mohsin M. Ghazal

Additional information is available at the end of the chapter

http://dx.doi.org/10.5772/intechopen.72483

Abstract

The samples of the mineralization of Pb-, Zn-, and Fe-sulfides were collected from three localities (Dure, Lefan, in the northern Thrust zone; and Sinjar, in the Foothill zone) in Northern Iraq. The geochemical recognition using X-ray diffraction (XRD) affirms the presence of the ore deposit sulfides (pyrite, sphalerite, galena, smithsonite, and cerussite). The characterization of mineral chemistry using electron microprobe analysis (EMPA) gives a clear and exact percentage of each element in each mineral. Fluid inclusions are mostly liquid H_2O and/or water vapor, which may also contain lesser soluble salts and slightly ore elements. Some fluid inclusions contain CO_2 vapor. This occurrence suggests the presence of two immiscible phases due to boiling at the time of their trapping. They are of epithermal system. The homogenization temperatures and salinities obtained for fluid inclusions can be comparable to those reported for the Mississippi Valley Type (MVT) lead-zinc deposits. It is concluded from the petrographic evidence, fluid inclusions and stable isotope data that lead-zinc mineralization was formed due to deeply circulating high-temperature fluids (brines) within the source basin, or later on by tectonic processes, which possibly contribute in leaching metals from either the diagenesis of host rocks or dewatering of deeper buried siliciclastic beds.

Keywords: mineralization, sulfides, Pb-Zn-Fe, geochemistry, paragenesis, Northern Iraq

1. Introduction

The Northern Iraqi Thrust zone and Foothill zone, where the three studied sections of the present study (Dure, Lefan, and Sinjar) are located, is considered as a good place of mineralization of many ore deposits, especially the lead-, zinc-, and iron-sulfides. This is because the structural, tectonic, and lithologic factors made essentially the highly fractured dolomitized limestones act as host rocks for this mineralization. Few detailed studies have been published which deal with the ore deposit mineralization in the area. First, Al-Bassam et al. [1] mentioned and studied the presence of lead and zinc sulfides. Then, other studies have been accomplished, but the more recent detailed geochemical investigation and exploration were carried out by Awadh [2, 3] and Shingaly [4]. The present chapter deals with the mineralization of sulfides in the above-mentioned three sections. It is worth to mention that this mineralization in Sinjar area is given for first time in such geochemical point of view. The work is a part of PhD research performed at Mosul University by the second author. The geochemical prospect of this work aims first to document the presence of the mineralization of the sulfide ore deposits in Northern Iraq, and second to elucidate their paragenesis, their source, and the processes responsible for their occurrence through the analysis of data obtained by the techniques of X-ray diffraction (XRD), electron microprobe analysis (EMPA), fluid inclusions, and stable isotopes of ^{13}C, ^{18}O, and ^{34}S.

2. Geology

The area of study had been influenced by Alpine orogeny that affected strongly the northern part of Iraq. During the time from Permian to mid-Cretaceous, the northern to eastern parts of the Arabian plate were subsiding gradually and bordering the Neo-Tethys Ocean [5]. The northern thrust zone on the Iraqi territories is formed to be as a developed ridge at the Arabian plate throughout the Zagros suture formed within the domain of the Neo-Tethys. During the movement of the Arabian plate and Tethys development, the mineralization of ore deposits occurred in the area as documented by Al-Bassam et al. [1], and later on by others as, for example, Awadh [3]. The host carbonate rocks of the ore deposits are represented mostly by fractures and karstified dolomitized limestones of Qamchuqa and Mergi formations (Middle Cretaceous), Bekhme formation (Upper Campanian), marly limestone of Shianish formation (Upper Campanian-Maastrichtian), and Kurra Chine formation (Upper Triassic) [6, 7].

The general geology and locations of the three studied sections (Dure, Lefan, and Sinjar) are illustrated in **Figure 1**.

3. Methodology

Representative samples were analyzed using a Philips X-ray diffractometer (PW3710) scanning from 4° to 60° 2θ. The generator was controlled using Philips PC-APD software. Peak identification was enabled using PDF/ICCD database and quantification using Rietveld analysis using commercial program Siroquant (Sietronics, Australia). Analysis was done at laboratories of the Department of Earth Sciences, Royal Holloway of London University.

Quantitative chemical analyses for selective host and sulfide minerals from the studied lead-zinc deposits were obtained from 10 polished sections (carbon coated) with a Cameca 3-spectrometer electron microprobe at University-College London UCL, UK. A defocused 15–25 μm beam was utilized at an accelerating voltage of 15 kV, a sample current of 15 nA, and a counting time of 10 s.

http://dx.doi.org/10.5772/intechopen.72483

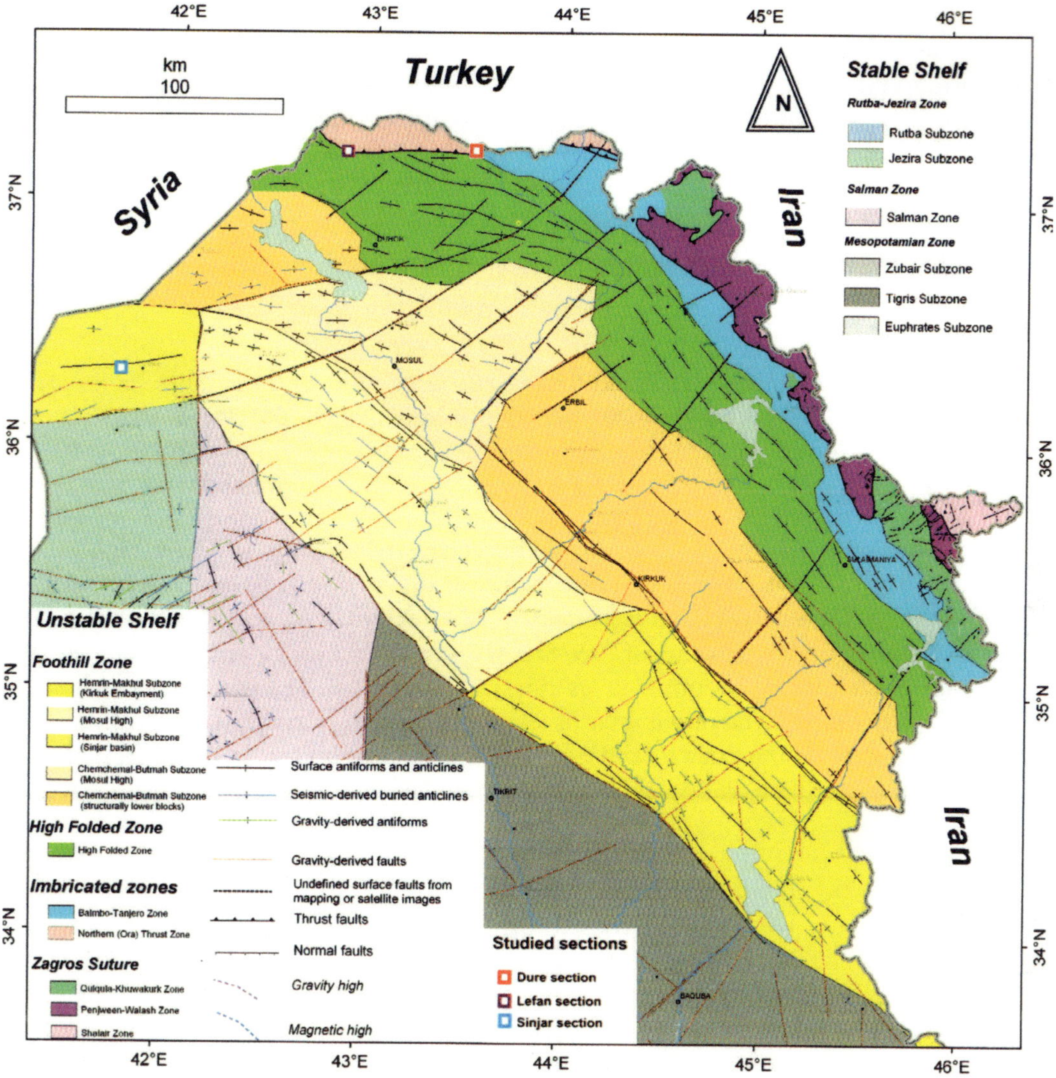

Figure 1. Geology and locations of the studied sections (after Jassim and Goff [8]).

Fluid inclusions were observed under petrographic microscope, and the following parameters were determined by microthermometry: highest temperature of ice melting (last ice melting temperature) (Tm) and homogenization/filling temperature (Th) measurements. Eight double-polished thin sections were prepared from the carbonates (dolomite and calcite) and sphalerite. The microthermometric measurements were taken on a NIKON Labophot-pol microscope mounted with LINKAM THMS-600 and TMS-92 freezing-heating stage and long-distance LW40x objective, in Iran Mineral Processing Research Center (IMPRC), and one sample was also carried out at Geological Engineering Department of Cumhuriyet University in Sivas/Turkey.

Mineral separates for sulfur isotope analysis were acquired by handpicking, and samples were converted to sulfur dioxide gas using a VG Isotech SIRA II mass spectrometer and using laboratory standard gas as a reference to produce true $\delta^{34}S$. The standards employed were the international standards NBS-123 and IAEA-S-3, and the SUERC standard CP-1. Analyses were made at an NERC Isotope Community Support Facility, SUERC, Glasgow, Scotland.

4. Geochemistry of ore deposits

4.1. Recognition

XRD diffractograms definitely illustrated the specific recognition of the sulfides (pyrite, FeS_2; sphalerite, ZnS; galena, PbS; smithsonite, $ZnCO_3$; and cerussite, $PbCO_3$) (**Figure 2**).

4.2. Characterization

The data provided in **Tables 1–3** reveal clearly the chemistry of pyrite, sphalerite, and galena, respectively. The pyrite in **Table 1** composes mainly of 44.7–47.5 wt% Fe and 46.6–55.1 wt% S. The sphalerite in **Table 2** composes essentially of 53.6–69.1 wt% Zn and 31–35.9 wt% S. The galena in **Table 3** composes mainly of 79.9–87.9 wt% Pb and 13.1–14.7 wt% S. Moreover, the calculations of their chemical formulae are given in **Tables 4–6**. The general paragenesis of pyrite, sphalerite, and galena of the three studied sections is illustrated in **Figure 3**. Pyrite is shown to be present in the three localities, where it is more abundant in both main and late stages at both Dure and Sinjar, while it is more abundant in both early and main stages at Lefan.

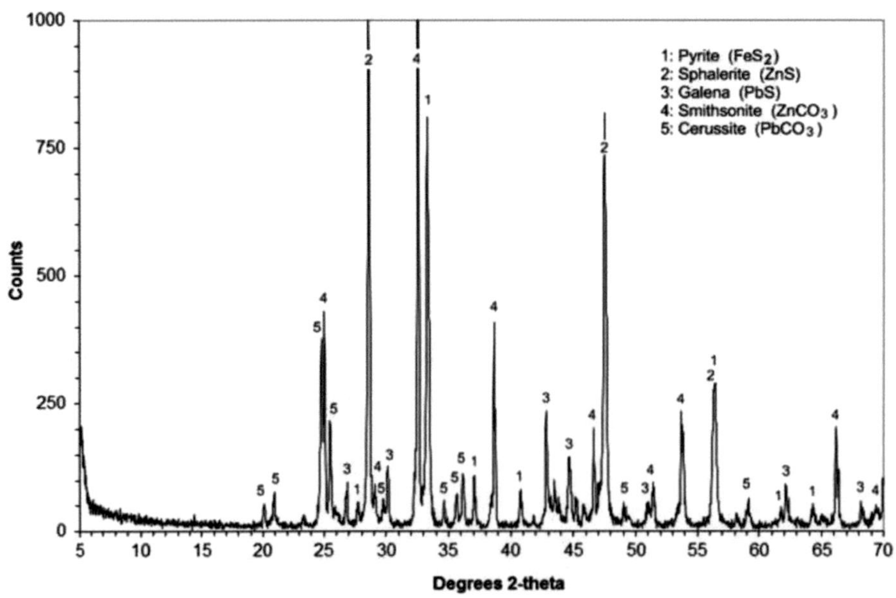

Figure 2. XRD diffractogram for recognition of ore deposits.

	Fe	S	As	Pb	Zn	Ni	Cu	Co	Total	Ore stage
DO 20-1	45.23	51.926	2.695	0.063	0.069	0.0001	0.078	0.003	100.0641	Early
DO 20-2	46.252	52.077	2.247	0.122	0.017	0.0001	0.032	0.003	100.7501	Early
DO 20-3	46.334	52.899	2.225	n.d	0.022	n.d	n.d	n.d	101.48	Early
DO 25-1	46.119	50.26	2.406	0.141	0.16	n.d	0.031	n.d	99.117	Early
DO 25-2	46.583	53.208	1.159	0.056	0.077	n.d	0.047	n.d	101.13	Early
DO 20-1	44.113	51.21	3.803	0.272	1.44	n.d	0.143	0.074	101.055	Main
DO 20-2	44.92	51.35	2.47	0.0931	1.96	0.0001	0.1015	0.003	100.8977	Main
DO 20-3	46.465	49.652	1.805	0.207	0.125	0.0001	0.062	0.002	98.3181	Main
DO 20-4	45.806	52.762	1.253	0.05	1.08	n.d	0.012	0.073	101.036	Main
DO 20-5	46.205	50.849	1.266	1.135	0.013	n.d	n.d	n.d	99.468	Main
DO 20-6	46.526	53.535	1.162	0.059	0.087	n.d	n.d	n.d	101.369	Main
DO 25-1	45.385	54.273	0.058	0.183	0.025	0.0001	0.087	0.1	100.1111	Main
DO 25-2	45.412	54.549	1.039	0.049	0.012	n.d	0.021	0.073	101.155	Main
DO 25-3	45.49	55.141	0.072	0.077	0.021	n.d	0.085	0.099	100.985	Main
DO 25-4	44.939	54.938	0.106	0.107	0.021	0.0001	0.016	0.063	100.1901	Main
DO 26-1	44.745	55.095	0.083	0.088	0.031	0.0001	0.057	0.051	100.1501	Main
DO 26-2	45.29	54.02	1.16	0.028	1.13	n.d	0.0038	n.d	101.6318	Main
DO 26-3	45.59	54.61	n.d	0.1115	1.25	n.d	0.1797	n.d	101.7412	Main
DO 26-4	46.16	53.73	n.d	0.0828	0.6338	n.d	n.d	n.d	100.6066	Main
DO 26-5	46.23	51.06	3.3	0.0489	0.3726	n.d	0.1521	n.d	101.1636	Main
DO 26-6	47.31	53.65	n.d	0.1165	0.232	n.d	0.0269	n.d	101.3354	Main
DO 25-1	47.04	53.75	0.238	0.0066	0.0135	n.d	0.4497	0.002	101.4998	Late
DO 25-2	47.48	52.97	0.0122	0.0673	n.d	n.d	0.0363	n.d	100.5658	Late
DO 25-3	46.22	54.12	0.1334	0.0351	0.6656	n.d	0.0149	0.003	101.192	Late
DO 25-4	46.67	46.66	0.265	0.1967	1.33	n.d	0.0429	n.d	95.1646	Late
DO 25-5	45.79	49.32	2.14	0.1811	1.33	n.d	0.054	n.d	98.8151	Late
DO 25-6	41.85	49.43	1.11	0.2752	5.83	n.d	0.0153	n.d	98.5105	Late
DO 25-7	43.77	50.73	1.96	1.64	1.86	n.d	0.0534	n.d	100.0134	Late
DO 25-8	45.7	49.35	1.89	1.49	2.15	n.d	0.0725	n.d	100.6525	Late
SO 4-1	47.31	54.81	0.003	0.02	0.004	0.04	0	0.002	102.189	Main
SO 4-2	46.12	55.17	0.004	0.004	0.035	0.03	0	n.d	101.363	Main

n.d = not detected.

Table 1. Results of electron microprobe analysis and paragenesis of pyrite in Dure section (DO) and Sinjar section (SO). Elements in wt%.

	Zn	S	Fe	Mn	Cd	Total	Ore stage
DO 26-1	61.11	33.51	5.14	0.016	0.221	99.997	Early
DO 26-2	61.66	33.25	4.23	0.001	0.213	99.354	Early
DO 26-3	61.16	33.2	6.078	0.008	0.227	100.673	Early
DO 26-1	59.81	34.35	6.524	0.011	0.196	100.891	Early
DO 30-1	61.62	32.75	5.514	0.002	0.229	100.115	Main
DO 20-1	62.64	33.53	3.993	0.004	0.236	100.403	Main
DO 20-2	63.2	33.83	3.608	n.d	0.263	100.901	Main
DO 20-3	64.03	32.61	3.797	0.018	0.257	100.712	Main
DO 20-4	63.88	32.52	3.651	n.d	0.226	100.277	Main
DO 25-1	62.39	34.88	2.33	0.003	0.225	99.828	Late
DO 25-2	65.34	34.4	0.81	n.d	0.263	100.813	Late
DO 25-3	65.49	33.55	1.82	n.d	0.277	101.137	Late
DO 25-4	64.49	33.01	2.43	n.d	0.311	100.241	Late
DO 25-5	64.56	33.04	2.62	n.d	0.334	100.554	Late
DO 25-6	69.11	31.01	0.177	n.d	0.244	100.541	Late
DO 25-7	66.83	33.05	0.706	n.d	0.286	100.872	Late
DO 25-8	65.09	33.95	0.968	0.004	0.255	100.267	Late
DO 25-9	65.12	33.69	1.19	0.005	0.26	100.265	Late
DO 30-1	66.22	33.12	0.812	n.d	0.265	100.417	Late
DO 30-2	67.49	32.29	0.129	0.025	0.235	100.169	Late
DO 30-3	66.33	34.08	0.088	n.d	0.279	100.777	Late
DO 30-4	66.29	34.36	0.113	n.d	0.307	101.07	Late
DO 30-5	67.11	33.18	0.052	0.007	0.216	100.565	Late
DO 30-6	68.45	32.31	0.255	0.005	0.263	101.283	Late
DO 30-7	67.65	33.1	0.437	n.d	0.237	101.424	Late
DO 30-9	65.94	33.15	0.259	0.007	0.263	99.619	Late
DO 30-10	66.79	33.28	0.051	0.021	0.272	100.414	Late
LO 10-1	54.5	35.01	9.52	0.001	0.002	99.033	Early
LO 10-2	53.88	35.11	10.14	0.05	0.09	99.27	Early
LO 10-3	53.61	35.91	10.56	0.05	0.09	100.22	Early
LO 18-1	59.1	33.56	7.4	n.d	0.02	100.08	Main
LO 18-2	59.05	33.45	7.45	n.d	0.05	100	Main
LO 18-3	59.41	33.34	7.46	n.d	0.09	100.3	Main
LO 18-4	58.91	33.38	7.82	n.d	0.5	100.61	Main
LO 18-1	62.86	33.51	3.47	0.005	0.5	100.345	Late

	Zn	S	Fe	Mn	Cd	Total	Ore stage
LO 18-2	62.84	33.24	3.49	0.01	0.11	99.69	Late
LO 18-3	62.71	33.56	3.57	n.d	0.19	100.03	Late
SO 4-1	65.1	35.7	0.27	n.d	0.15	101.22	Main

n.d = not detected.

Table 2. Results of electron microprobe analysis and paragenesis of sphalerite in Dure section (DO), Lefan section (LO) and Sinjar section (SO). Elements in wt%.

Sphalerite at Dure is more abundant in the main stage, while at Lefan it is more in both main and late stages. In Sinjar it is recorded only in the late stage with few abundance.

Galena is found more abundant in late stage and less in the main stage at Dure, while it is more in the main stage than in the late stage at Lefan, and no galena is recorded at Sinjar.

Pyrite varies in composition as shown in **Table 4**. At Dure, the pyrite of the early stage has mainly arsenic as traces in Fe-site ($As_{.02-.042}$); other lesser traces are of zinc and copper ($Zn_{.001}$; $Cu_{.001}$). In the main stage, pyrite has relatively more arsenic reaching up to $As_{.06}$, while other

	Pb	S	Fe	Zn	Cd	Total	Ore stage
DO 26-1	86.403	13.612	0.022	n.d	n.d	100.037	Main
DO 26-2	86.635	13.843	0.016	n.d	n.d	100.494	Main
DO 26-3	86.443	13.755	0.018	0.07	n.d	100.286	Main
DO 26-4	87.934	13.154	0.003	0.022	n.d	101.113	Main
DO 25-1	84.73	13.86	0.4338	3.63	0.1672	102.821	Late
DO 25-2	82.75	14.69	0.1514	3.99	0.2513	101.8327	Late
DO 25-3	82.16	14.19	0.3233	3.88	0.1941	100.7474	Late
DO 25-4	82.92	13.86	0.2788	3.15	0.2013	100.4101	Late
DO 25-5	82.52	13.89	0.2209	3.45	0.1777	100.2586	Late
DO 25-6	79.94	13.61	0.7983	5.5	0.1723	100.0206	Late
DO 25-7	86.21	13.86	0.1214	1.36	0.2248	101.7762	Late
DO 25-8	83.11	13.96	0.112	2.13	0.1897	99.5017	Late
DO 25-9	83.16	13.43	0.0881	4.91	0.2263	101.8144	Late
LO 10-1	85.1	13.21	0.04	0.33	n.d	98.68	Main
LO 10-2	84.6	13.22	0.02	0.31	n.d	98.15	Main
LO 10-3	84.03	13.11	n.d	2.34	0.18	99.66	Late

n.d = not detected.

Table 3. Results of electron microprobe analysis and paragenesis of galena in Dure section (DO) and Lefan section (LO). Elements in wt%.

Sample no.	Chemical formula	Ore stage
DO 20-1	Fe(.955 As.042 Zn.001 Cu.001) S2	Early
DO 20-2	Fe(.963 As.035 Cu.001) S2	Early
DO 20-3	Fe(.965 As.035) S2	Early
DO 25-1	Fe(.96 As.04) S2	Early
DO 25-2	Fe(.98 As.02) S2	Early
DO 20-1	Fe(.91 As.06 Pb.002 Zn.03 Cu.003 Co.001) S2	Main
DO 20-2	Fe(.93 As.04 Zn.03) S2	Main
DO 20-3	Fe(.97 As.03) S2	Main
DO 20-4	Fe(.96 As.02 Zn.02) S2	Main
DO 20-5	Fe(.97 As.02 Pb.01) S2	Main
DO 20-6	Fe(.98 As.02) S2	Main
DO 25-1	Fe(.99 As.001 Pb.001 Zn.001 Cu.002 Co.002) S2	Main
DO 25-2	Fe(.98 As.02) S2	Main
DO 25-3	Fe(.99 As.001 Cu.002 Co.002) S2	Main
DO 25-4	Fe(.995 As.002 Pb.001 Co.001) S2	Main
DO 26-1	Fe(.995 As.001 Pb.001 Zn.001 Co.001) S2	Main
DO 26-2	Fe(.96 As.02 Zn.02) S2	Main
DO 26-3	Fe(.973 Pb.001 Zn.023 Cu.003) S2	Main
DO 26-4	Fe(.99 Zn.01) S2	Main
DO 26-5	Fe(.94 As.05 Zn.007 Cu.003) S2	Main
DO 26-6	Fe(.995 Pb..001 Zn.004) S2	Main
DO 25-1	Fe(.988 As.004 Cu.008) S2	Late
DO 25-2	Fe(.999 Cu.001) S2	Late
DO 25-3	Fe(.985 As.002 Zn.012) S2	Late
DO 25-4	Fe(.97 As.004 Pb.001 Zn.024 Cu.001) S2	Late
DO 25-5	Fe(.942 As.033 Pb.001 Zn.023 Cu.001) S2	Late
DO 25-6	Fe(.877 As.017 Pb.002 Zn.104) S2	Late
DO 25-7	Fe(.925 As.031 Pb.009 Zn.034 Cu.001) S2	Late
DO 25-8	Fe(.925 As.029 Pb.008 Zn.037 Cu.001) S2	Late
SO 4-1	FeS2	Main
SO 4-2	Fe(.999 Zn.001) S2	Main

Table 4. Chemical formula of pyrite (calculated from data in **Table 1**).

traces ($Zn_{.001-.104}$, $Pb_{.001}$ $Cu_{.002-.008}$ and $Co_{.001-.002}$) may be found .In the late stage, pyrite has no difference in arsenic content than that of the early stage ($As_{.002-.04}$). The pyrite at Sinjar is very pure (FeS_2); only $Zn_{.001}$ is recorded. At Lefan, no data are available.

http://dx.doi.org/10.5772/intechopen.72483

Sample no.	Chemical formula	Ore stage
DO 26-1	Zn(.908 Fe.089 Cd.002) S	Early
DO 26-2	Zn(.923 Fe.074 Cd.002) S	Early
DO 26-3	Zn(.894 Fe.104 Cd.002) S	Early
DO 26-1	Zn(.885 Fe.113 Cd.002) S	Early
DO 30-1	Zn(.903 Fe.095 Cd.002) S	Main
DO 20-1	Zn(.928 Fe.069 Cd.002) S	Main
DO 20-2	Zn(.935 Fe.063 Cd.002) S	Main
DO 20-3	Zn(.933 Fe.065 Cd.002) S	Main
DO 20-4	Zn(.935 Fe.063 Cd.002) S	Main
DO 25-1	Zn(.956 Fe.042 Cd.002) S	Late
DO 25-2	Zn(.983 Fe.014 Cd.002) S	Late
DO 25-3	Zn(.966 Fe.031 Cd.002) S	Late
DO 25-4	Zn(.955 Fe.042 Cd.003) S	Late
DO 25-5	Zn(.952 Fe.045 Cd.003) S	Late
DO 25-6	Zn(.995 Fe.003 Cd.002) S	Late
DO 25-7	Zn(.985 Fe.012 Cd.003) S	Late
DO 25-8	Zn(.981 Fe.017 Cd.002) S	Late
DO 25-9	Zn(.977 Fe.021 Cd.002) S	Late
DO 30-1	Zn(.984 Fe.014 Cd.002) S	Late
DO 30-2	Zn(.995 Fe.002 Cd.002) S	Late
DO 30-3	Zn(.996 Fe.002 Cd.002) S	Late
DO 30-4	Zn(.995 Fe.002 Cd.003) S	Late
DO 30-5	Zn(.997 Fe.001 Cd.002) S	Late
DO 30-6	Zn(.993 Fe.004 Cd.002) S	Late
DO 30-7	Zn(.99 Fe.008 Cd.002) S	Late
DO 30-9	Zn(.993 Fe.005Cd.002) S	Late
DO 30-10	Zn(.996 Fe.001 Cd.002) S	Late
LO 10-1	Zn(.83 Fe.17) S	Early
LO 10-2	Zn(.818 Fe.18 Mn.001 Cd.001) S	Early
LO 10-3	Zn(.811 Fe.187 Mn.001 Cd.001) S	Early
LO 18-1	Zn(.87 Fe.13) S	Main
LO 18-2	Zn(.87 Fe.13) S	Main
LO 18-3	Zn(.871 Fe.128 Cd.001) S	Main
LO 18-4	Zn(.862 Fe.134 Cd.004) S	Main
LO 18-1	Zn(.935 Fe.061 Cd.004) S	Late
LO 18-2	Zn(.938 Fe.061 Cd.001) S	Late

Sample no.	Chemical formula	Ore stage
LO 18-3	Zn(.936 Fe.062 Cd.002) S	Late
SO 4-1	Zn(.994 Fe.005 Cd.001) S	Main

Table 5. Chemical formula of sphalerite (calculated from data in **Table 2**).

Sphalerite as shown in **Table 5** is also noticed to have traces of $Fe_{.001–.095}$ and $Cd_{.002–.003}$ in the Zn site of its chemical formula of all stages at Dure, whereas at Lefan, sphalerite has sometimes only $Fe_{.13–.17}$ in its formula in both early and main stages, which is more than that at Dure; in other samples, there is $Cd_{.001–.004}$ that is recorded in all stages. At Sinjar, sphalerite has also $Fe_{.005}$ and $Cd_{.001}$ in the Zn site.

Sample no.	Chemical formula	Ore stage
DO 26-1	Pb(.999 Fe.001) S	Main
DO 26-2	Pb(.999 Fe.001) S	Main
DO 26-3	Pb(.996 Fe.001 Zn.003) S	Main
DO 26-4	Pb(.999 Zn.001) S	Main
DO 25-1	Pb(.863Fe .016 Zn.117 Cd.003) S	Late
DO 25-2	Pb(.858 Fe.006 Zn.131 Cd.005) S	Late
DO 25-3	Pb(.856 Fe.012 Zn.128 Cd.004) S	Late
DO 25-4	Pb(.879 Fe.011 Zn.106 Cd.004) S	Late
DO 25-5	Pb(.872 Fe.009 Zn.116 Cd.003) S	Late
DO 25-6	Pb(.794 Fe.029 Zn.173 Cd.003) S	Late
DO 25-7	Pb(.943 Fe.005 Zn.047 Cd.005) S	Late
DO 25-8	Pb(.917 Fe.005 Zn.074Cd.004) S	Late
DO 25-9	Pb(.836 Fe.003 Zn.156 Cd.004) S	Late
LO 10-1	Pb(.986 Fe.002 Zn.012) S	Main
LO 10-2	Pb(.988 Fe.001 Zn.011) S	Main
LO 10-3	Pb(.916 Zn.081 Cd.004) S	Late

Table 6. Chemical formula of galena (calculated from data in **Table 3**).

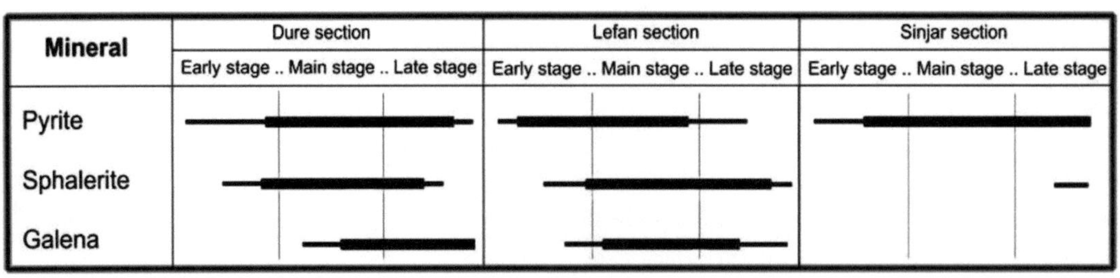

Figure 3. General paragenetic sequence of the sulfide ore deposits in the studied sections (Dure, Lefan, and Sinjar).

Galena in **Table 5** reveals also some impurities as traces in Pb site. At Dure, $Fe_{.001}$ and $Zn_{.001-.003}$ in the main stage are recorded in pyrite, while in the late stage, $Fe_{.003-.29}$, $Zn_{.047-.173}$, and $Cd_{.003-.005}$ as traces in Pb site of pyrite are recorded.

At Lefan, galena has traces of $Zn_{.011-.081}$, $Fe_{.001-.002}$, and $Cd_{.004}$ in the only main stage. No galena is recorded at Sinjar.

5. Paragenesis

Generally, fluid inclusion and stable isotope data constrain the different stages of the rock evolution [9].

5.1. Fluid inclusions

Most fluid inclusions are essentially composed of a liquid phase and/or a vapor bubble, but they may also contain soluble salts and slightly ore elements.

In **Table 7**, the homogenization temperatures are measured for 22 fluid inclusions in sphalerite, dolomite, and calcite minerals from the Dure deposits and 13 fluid inclusions in dolomite and calcite minerals from the Lefan deposits are given. This is helpful to constrain the hydrothermal fluid conditions during the formation of these deposits and to find out evidence of boiling. No suitable sample has been examined from Sinjar deposit for analysis [10].

The Dure deposit contains two phases of fluid inclusions at room temperature. They are H_2O-rich liquid with H_2O vapor bubbles constituting typically 5–10% vapor volume.

Primary fluid inclusions occur as single or clustered rounded or irregular shape inclusions that are ranging in size from 5 μm to < 10 μm (**Figure 4A**). Negative crystal forms are uncommon. Few primary inclusions have vapor volumes greater than 15% (**Figure 4A**). Secondary fluid inclusions occur as fracture-controlled planar groups, and they are thin, elongated, and irregular in shape (**Figure 4B**). The inclusions are <5 μm in size.

At room temperature, the Lefan deposit contains liquid-rich and vapor-rich fluid inclusions. The fluid inclusions typically contain H_2O-rich liquid with H_2O vapor bubbles. Some fluid inclusions contain CO_2 vapor. Vapor-rich inclusions are subrounded or irregularly shaped and occur as clusters containing one to three inclusions (**Figure 4C**). These inclusions are larger in size, 5–12 μm, than the liquid-rich inclusions, <5μm.

Primary fluid inclusions are irregular or rounded and randomly distributed in carbonate minerals (**Figure 4D**). Secondary fluid inclusions form as planar groups of elongated inclusions (**Figure 4E**). The presence of fluid inclusions with variable liquid to vapor volumetric phase ratios is the most common evidence of entrapment from boiling fluids [11]. The vapor-rich inclusions are believed to result from boiling fluids rather than necking-down processes (**Figure 4F**) due to their large size relative to the liquid-rich inclusions. There are absence of one-phase liquid inclusions and the presence of fluid-rich inclusions with relatively constant liquid:vapor phase ratios and consistent Th. The occurrence of vapor-rich and liquid-rich

Sample no.	T_m (°C) Ice melting temperature	T_h (°C) Homogenization temperature	Salinity wt% NaCl equivalent	Origin	Mineral
DO 35					
1	−14.1	174	17.79	Primary	Sphalerite
2	−14.3	173	17.96	Primary	Sphalerite
3	−13.8	183	17.61	Primary	Sphalerite
4		180		Secondary	Sphalerite
5		177		Secondary	Sphalerite
6		179		Secondary	Sphalerite
7		165		Secondary	Sphalerite
8		157		Secondary	Sphalerite
DO 28					
1	−29.1	144	>23	Primary	Saddle dolomite
2	−23.2	147	>23	Primary	Saddle dolomite
3	−18.1	128	20.97	Primary	Saddle dolomite
4		115		Secondary	Saddle dolomite
5		117		Secondary	Saddle dolomite
6		97		Secondary	Saddle dolomite
7		127		Secondary	Saddle dolomite
LO 15					
1	−5.3	284	8.14	Primary	Saddle dolomite
2	−3.2	200	5.11	Primary	Saddle dolomite
2	−5.3	196	8.14	Primary	Saddle dolomite
4	−6.9	210	10.36	Primary	Saddle dolomite
5	−2.2	238	3.55	Primary	Saddle dolomite
6	−1.8	225	3.06	Primary	Saddle dolomite
7		193		Secondary	Saddle dolomite
8		196		Secondary	Saddle dolomite
DO 10					
1	−10.1	61	13.93	Primary	Calcite vein
2	−33.2	59	>23	Primary	Calcite vein
3	−35.1	105	>23	Primary	Calcite vein
4		45		Secondary	Calcite vein
5		50		Secondary	Calcite vein
6		65		Secondary	Calcite vein
7		74		Secondary	Calcite vein

Sample no.	T_m (°C) Ice melting temperature	T_h (°C) Homogenization temperature	Salinity wt% NaCl equivalent	Origin	Mineral
LO 10					
1	−7.8	127	11.46	Primary	Calcite vein
2	−10.2	145	10.04	Primary	Calcite vein
3	−10.6	157	14.57	Primary	Calcite vein
4		71		Secondary	Calcite vein
5		68		Secondary	Calcite vein

Table 7. Homogenization temperatures and salinities of fluid inclusions in sphalerite and carbonate minerals from the Dure (DO-sample) and Lefan (LO-sample) deposits.

inclusions suggests the presence of two immiscible phases due to boiling at the time that the fluid inclusions were trapped. Textures indicative of boiling are present in the Lefan deposit, but are not as common as those cited in other epithermal systems (e.g., Dure section).

5.2. Microthermometric analysis

Microthermometric analyses have been carried out on fluid inclusions in sphalerite, dolomites, and calcites to obtain a preliminary estimate of the temperatures and salinities of the ore-bearing fluid. The results of the microthermometric analyses are summarized in **Table 7**. Suitable fluid inclusions are observed only in saddle dolomite and calcite. Few suitable fluid inclusions are identified in sphalerite, where the observed inclusions are either too small for microthermometric measurement or they are judged to be secondary (**Figure 4A**).

Most inclusions chosen for analyses are considered as primary (contemporaneous with their host minerals) and closely related spatially and genetically to the sulfide mineralization. These inclusions are generally small (about 5 μm) and contain no daughter minerals (**Figure 4A, C,** and **D**). Due to the small size of the inclusions, proving accurate geothermometry is difficult. In addition, some hydrocarbon-rich fluid inclusions are also found in gangue minerals. Fluid inclusions in the late-generated barite within Bekhme formation at Lefan lead-zinc deposit are considered as fingerprint of hydrocarbon generation [3].

Salinities of fluid inclusions are determined by measuring the freezing temperatures of the fluid inclusions. According to Bodnar and Vityk [12], the freezing point of a fluid inclusion is the temperature at which the last ice crystal melts (ice melting temperature). The equation used to determine the salinity of a fluid inclusion is based on the H_2O-NaCI system following Bodnar and Vityk [12].

$$\text{Salinity} = 0.00 + 1.78\theta – 0.0442\theta2 + 0.000557\theta3 \quad (1)$$

where salinity = the weight percent (wt %) NaCl in solution and θ = freezing point depression (ice melting temperature) in °C. Freezing temperatures were measured prior to homogenization temperatures to avoid leakage or decrepitation of the fluid inclusions [13].

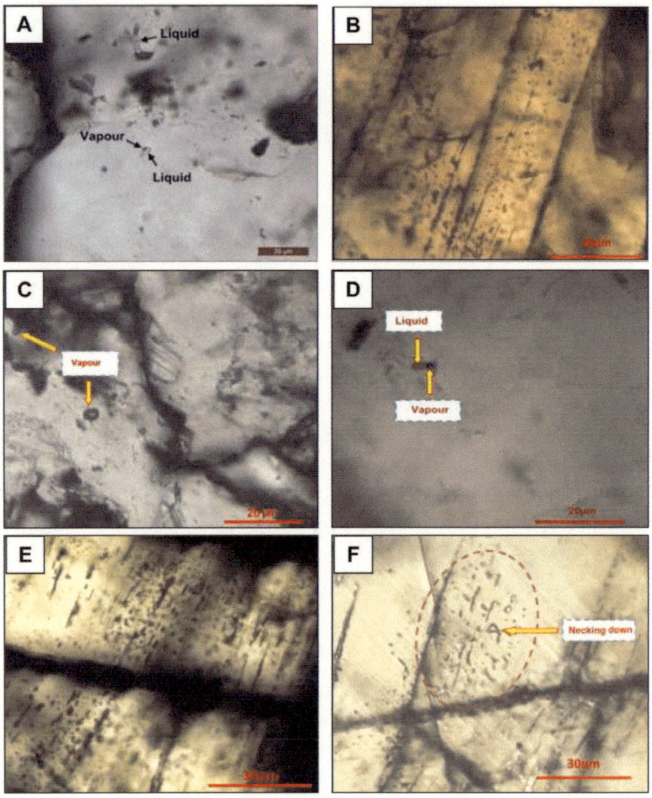

Figure 4. Primary and secondary fluid inclusions from Dure and Lefan lead-zinc deposits. (A) Primary 2-phase fluid inclusion and (B) secondary fluid inclusions within sphalerite (DO 35, Dure section). (C) Primary single-phase vapor inclusions within saddle dolomite (LO 15, Lefan section). (D) Primary two-phase fluid inclusion within calcite (LO 10, Lefan section). (E) Elongated-secondary fluid inclusions within saddle dolomite (LO 15, Lefan section). (F) Necking-down process within saddle dolomite (LO 15, Lefan section). Numbers on scale bars in microns.

The salinities of the fluid inclusions in Dure deposit range from 13.93 to >23 wt% NaCl equivalent, whereas in the Lefan deposit they range from 3.06 to 14.57 wt% NaCl equivalent (**Table** 7).

Homogenization temperatures were determined for primary and secondary inclusions with constant liquid:vapor volumetric phase ratios to avoid inaccurate measurements that could result from necking-down processes, leakage, stretching and decrepitation [11]. Homogenization temperatures of the fluid inclusions in Dure deposit range from 45°C to 183°C, whereas the Lefan deposit has fluid inclusions that homogenize at temperatures ranging from 68°C to 284°C (**Table** 7).

Both Dure and Lefan deposits have relatively some differences in fluid inclusion homogenization temperatures. The primary fluid inclusions in the Dure and Lefan deposits yielded homogenization temperatures with an average value of 130.4°C and 198°C, respectively, although values range from approximately 60°C to 284°C for both sections (**Table** 7).

5.3. Interpretation of fluid inclusion results

The homogenization temperatures and salinities obtained for fluid inclusions in the present study are nearly similar to those reported for lead-zinc deposits of Mississippi Valley Type (MVT). Many authors such as Leach and Sangster [14], Paradis et al. [15], and Leach et al. [16, 17] have given the fluid inclusion temperatures in MVT deposits to range from about 50°C to 250°C; however, most of the measured temperatures were between 75°C and 200°C. The salinities of MVT fluids determined by those authors from fluid inclusions are typically 10–30 wt.% NaCl equivalent.

Both primary and secondary fluid inclusions of the deposits at Dure area contain the lowest Th values when compared to those of the Lefan area. The Lefan deposits have some vapor-rich fluid inclusions homogenized at relatively higher temperatures (200–280°C). These anomalous high temperatures can probably attributed to a vapor-rich fluid that trapped some liquid during boiling. During crystallization of sphalerite, the high temperature was preceded, but relatively moderate formation temperatures during the formation of saddle dolomite followed by cooling to about 50°C for late calcite formation. This supports the hypothesis of dilution and cooling of the transmitted fluids in the cracks and fissures of the host rocks by mixing with meteoric waters during later formed sulfides or perhaps post-sulfide calcite stages of deposition.

The diagram of salinity versus homogenization temperature of the fluid inclusions in the studied sections reveals no limit relationship (**Figure 5**), but rather scattered between inclusion salinities and homogenization temperatures. This may be due to mixing of two different solutions; one was hot and saline, and the other was cool with low salinity (e.g., Taylor et al. [18]) as in Lefan section. Furthermore, invariant salinities of inclusions are interpreted as evidence of precipitation from a single fluid [19], as in Dure section.

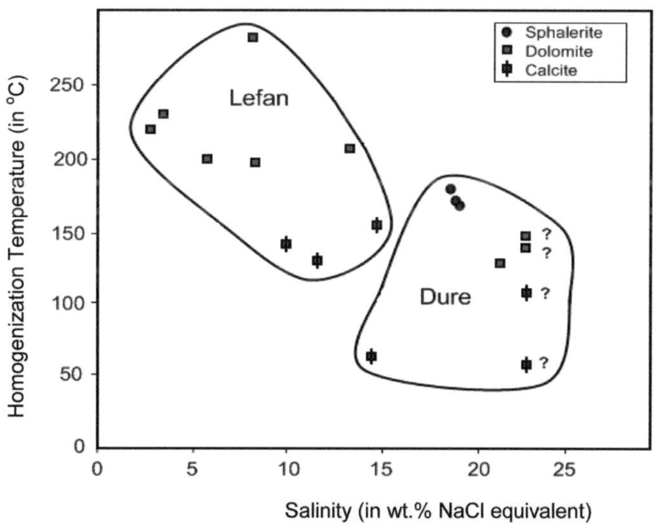

Figure 5. Fluid inclusion salinity vs. homogenization temperature for the studied deposits (Dure and Lefan sections). Question mark means that the absolute value of salinity is not determined (i.e., >23%).

Accordingly, the mineralization of the studied sulfides is seen to follow the type of MVT mineralization. However, the similarities between MVT inclusion fluids and oil-field brines are established to a wide acceptance of a basin-generated origin for MVT fluids. The high salinity is explained by the dissolution of evaporites, incorporation of connate bittern brines, or through infiltration of evaporated surface waters of the sedimentary basin brines [16, 17, 20].

5.4. Genesis of sulfides from stable isotopes

The average values of $\delta^{34}S$‰ of the studied sulfide of Dure, Lefan, and Sinjar are (0.8, 1.8, and 30.8‰), respectively. Sulfur isotope analysis shows that the sulfur in Dure and Lefan is originated from a mixture of various sources, but probably derived from seawater dissolved sulfate and/or diagenetic pyrite for Sinjar deposits. The $\delta^{13}C$ and $\delta^{18}O$ values of carbonate-host rocks are in the range of marine carbonates. Petrographic evidence and stable isotope data with fluid inclusions suggest that lead-zinc mineralization was caused by deeply circulating mineralizing fluids of high temperature (brines) within the source basin of deposition or due to tectonic processes, which possibly contribute in leaching metals from either the diagenetic host rocks or dewatering of deeper buried siliciclastic beds. The studied carbonate-hosted lead-zinc deposits seem to have many similarities with Mississippi Valley Type (MVT) deposit.

Deposit locality	Sample	Mineral	$\delta^{34}S$ ‰	Author(s)
Dure	DO 25	Pyrite	0.9	Present study
	DO 22	Pyrite	1.3	Present study
	DO 15	Pyrite	1.6	Present study
		Pyrite	0.2	Al-Bassam et al. [1]
		Pyrite	3.6	Al-Bassam et al. [1]
	DO 22	Sphalerite	2.4	Present study
	DO 30	Sphalerite	2.0	Present study
		Sphalerite	0.0	Al-Bassam et al. [1]
		Sphalerite	−0.4	Al-Bassam et al. [1]
	DO 28	Galena	1.8	Present study
	DO 30	Galena	−0.9	Present study
		Galena	−1.8	Al-Bassam et al. [1]
		Galena	−2.6	Al-Bassam et al. [1]
Lefan	LO 22	Pyrite	3.6	Present study
	LO 15	Sphalerite	1.8	Present study
	LO 15	Galena	1.2	Present study
	LO 18	Galena	0.47	Present study

Table 8. Sulfur isotope ($\delta^{34}S$) values of the studied lead-zinc deposits. Data of pyrite (n = 2), sphalerite (n = 2), and galena (n = 2) in Dure deposits from Al-Bassam et al. [1] were also displayed.

The values of sulfur isotope in the Dure (average = 0.8‰) and Lefan (average = 1.8‰) (**Table 8**) cannot be attributed necessarily to a magmatic source of sulfur because the field observations together with ore microscopy provide no magmatic activity. The negative shift of $\delta^{34}S$ value is resulted from seawater sulfates, or bacterial reduction, but possibly some sulfur could be derived due to the dissolution and leaching of preexisting sulfide-bearing igneous rocks.

The positive $\delta^{34}S$ values suggest that sulfur was derived from seawater or ancient evaporates and connate water undergone subsequent reduction by ferrous iron or organic compounds [21] (see **Figure 6** for more details). Perhaps seawater became enriched in light isotope by leached sulfur from deeper lithologic units during water circulation [23–25]. The influence of the enrichment of light sulfur isotopes is more significant for the Dure deposits (average, 0.8‰) than that for Lefan (average, 1.8‰).

Previous studies [1, 2, 26], together with this study, assumed that the studied Pb-Zn deposits are epigenetic of sedimentary origin formed after lithification of their host rocks.

Several evidences suggest the basinal brines source for hydrothermal fluids in the studied sections that are responsible for ore formation: first, the salinities and temperatures of the fluid inclusions of these deposits are mainly within the range of fluids of the MVT deposits [16, 17]; Second, the similarity in composition between the ore deposits and the most basinal brines. Low and narrow variation in $\delta^{34}S$ values of sulfides (**Table 9**) may indicate a single source of sulfur. Because of the absence of magmatic activity, it is realistic to consider the rocks in the basin of the studied sections as the source of sulfur. This leads to accept the positive $\delta^{34}S$, the organic sulfur enrichment in the beds of slightly positive to negative $\delta^{34}S$, and the presence of diagenetic pyrite in sedimentary rocks. Third, carbon and oxygen isotope values in the carbonate-host rocks are generally of relatively lower $\delta^{18}O$ and higher $\delta^{13}C$. These values

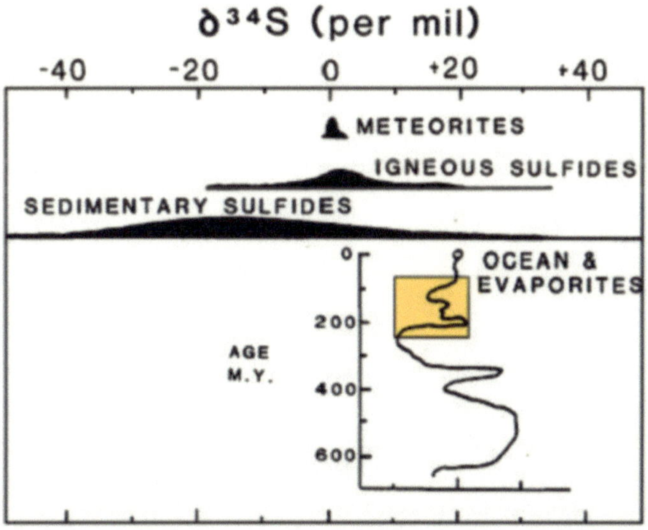

Figure 6. Sulfur isotopic variation in nature (from Ohmoto and Rye [22]).

Sample	Lithology	$\delta^{13}C_{PDB}$ (‰)	$\delta^{18}O_{PDB}$ (‰)	$\delta^{18}O_{SMOW}$ (‰)
DK 90	Dolomitized lime-mudstone	−1.05	−4.48	26.24
DK 84	Black dolomitic shale	0.60	−8.48	22.12
DK 80*	Fine dolomitized limestone	1.23	−7.71	22.91
DK 70*	Brecciated dolostone	1.16	−6.94	23.71
DK 65*	Saddle dolomite	−1.25	−11.93	18.56
DK 58*	Smithsonite ($ZnCO_3$)	−2.77	−10.09	20.46
DK 45	Coarse dolomitized limestone	0.59	−5.49	25.20
DK 32	Coarse gray dolostone	−1.57	−9.01	21.57
DK 25	Bioclastic lime-mudstone	−1.30	−8.70	21.89
DK 18	Brecciated dolomitic limestone	−2.80	−8.48	22.12
DK 10	Bioclastic lime-wackstone	−1.00	−9.01	25.01
LSh 1	Packstone (limestone)	0.31	−4.53	26.19
LB 68	Coarse dolostone	1.93	−8.19	22.42
LB 64	Dolomitized lime-wackstone	−0.41	−5.67	25.01
LB 60	Dolomitized lime-wackstone	2.23	−4.83	25.88
LB 55	Dolomitized lime-wackstone/packstone	1.37	−5.37	25.32
LB 53*	Dolostone with saddle dolomite	−2.53	−7.04	23.61
LB 35*	Dolomitized lime-packstone	1.61	−5.20	25.50
LB 32*	Dolomitized lime-grainstone	0.96	−6.15	24.52
LB 31*	Recrystallized limestone	0.94	−5.93	24.75
LB 28	Dolomitized lime-grainstone	−3.34	−7.13	23.51
LB 25	Dolomitic limestone	1.28	−3.77	26.98
LB 20	Recrystallized lime-wackstone	1.13	−5.91	24.77
LB 17	Dedolomitized dolostone	1.24	−4.87	25.84
LB 12	Rudist boundstone (limestone)	1.27	−3.94	26.80

DK samples from Dure section and LB samples from Lefan section. LSh 1 is one sample from Shiranish formation from Dure section.
*Mineralized samples. Others unmineralized samples.

Table 9. Carbon ($\delta^{13}C$) and oxygen ($\delta^{18}O$) isotope values of carbonate-host rocks.

require involvement of marine carbonates and organic matter in the basin to supply such carbon and oxygen (**Figure 7**). Moreover, there is a fact about the presence of hydrocarbon-rich fluid inclusions in some gangue minerals, and a late generation of barite [3], all supporting the probable basinal brines as the source of ore-forming fluids. As compared with typical basinal brines, some fluid inclusions in the present studied samples exhibit relatively lower salinities (<10 wt.% NaCl equiv.). The low salinity of these fluid inclusions shows a positive correlation with their Th content, possibly due to mixing between meteoric water and basinal brines.

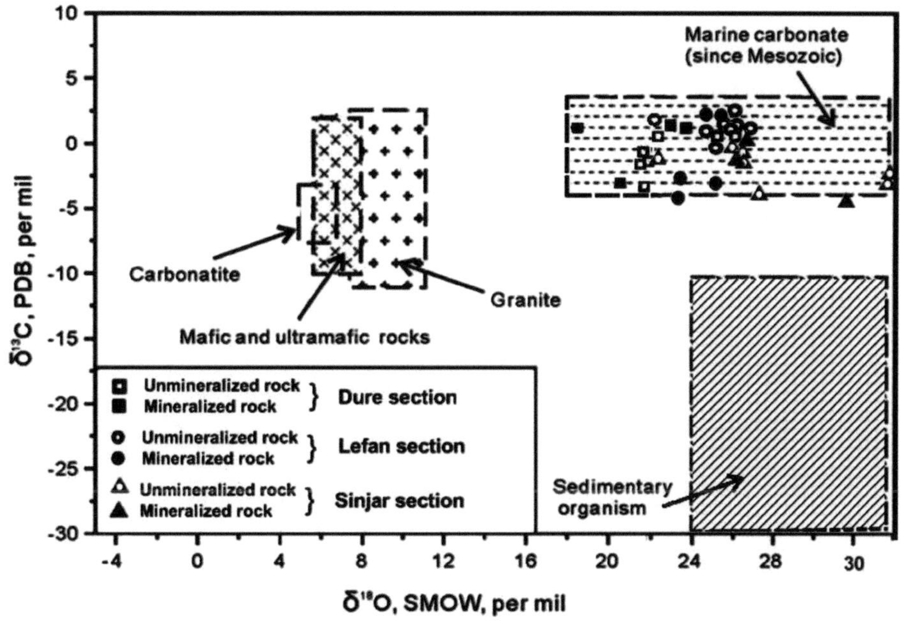

Figure 7. Plot of $\delta^{13}C$ vs. $\delta^{18}O$ of carbonate-host rocks of the studied deposits in the Dure and Lefan sections, showing $\delta^{13}C$ and $\delta^{18}O$ values of the unmineralized and mineralized carbonate rocks in the sections [4]. The fields of carbonatite, mafic and ultramafic rocks, granite, marine carbonate, and sedimentary organisms are after He et al. [27].

6. Conclusions

The geochemical prospect throughout the recognition, characterization, fluid inclusions, and stable isotopes of the studied ore deposits carried out in this study emphasizes the suggestion that the Pb-Zn-Fe mineralization is concentrated from brines (marine waters). It is thought that the circulation of hydrothermal water happened at depths, where cracks and faults were formed at the beginning of the orogeny event. At that depth, the fluid was heated enough to dissolve metals from their hosting rocks during its ascent, and as the transported metal ions concentrated sufficiently, deposition took place. This process, together with permeability of the wall rock, is thought to be the eventual controlling factor of mineralization and the present geometry of the ore bodies and the alteration. Mineralization tends to occur in fractures and solution-collapse features formed during uplift of platform-carbonate sequences accompanying plate convergence. Therefore, the preferred genetic model for the concentration of ore metals at the studied carbonate rocks should involve basin dewatering due to compaction and regional tectonism and expulsion of the basin-derived fluids into the brecciated, karstificated, and highly porous recrystallized and dolomitized host rocks of the Kurra Chine and Bekhme formations.

Acknowledgements

The authors would like to thank the Earth Science Department, College of Science, Mosul University, for supplying available facilities and administrative support. A special thank goes to Dr Dave Alderton from Royal Holloway, London University, for his help in XRD; Prof. Dr Ahmet Gokce, Cumhuriyet University, Turkey; and Mrs Soheila Aghajani, Iran Mineral Processing Research Center (IMPRC) for their contribution to the fluid inclusion analysis. Considerable gratitude goes to NERC Isotope Community Support Facility, SUERC, Glasgow, Scotland, for sulfur isotope analysis, and to University-College London UCL, UK, for support in electron microprobe analysis (EMPA).

Author details

Ali Ismail Al-Juboury[1]*, Waleed S. Shingaly[2], Elias M. Elias[1] and Mohsin M. Ghazal[1]

*Address all correspondence to: alialjubory@yahoo.com

1 Geology Department, Mosul University, Mosul, Iraq

2 Geology Department, Salahaddin University, Erbil, Iraq

References

[1] Al-Bassam K, Hak J, Watkinson D. Contribution to the origin of the Serguza lead-zinc-pyrite deposit, Northern Iraq, Mineral. Deposita. 1982;**17**:133-149

[2] Awadh SM. Mineralogy, geochemistry and origin of the zinc–lead–barite deposits from selected areas from north of Zakho, Northern Iraq [thesis]. Iraq: University of Baghdad; 2006

[3] Awadh SM, Al-Ameri TK, Jassim SY, Bayraktutan MS. Fluid inclusions usage for assessing oil migration in Duhok, north Iraq. Positioning. 2010;**1**:42-49

[4] Shingaly WSA. The role of carbonate host rocks in the genesis of lead-zinc sulfide deposits at Dure and Lefan sections and iron-sulfide deposit at Sinjar Section-Northern Iraq [thesis]. Iraq: Mosul University; 2013

[5] Sharland PR, Archer R, Casey D, et al. Arabian Plate Sequence Stratigrpahy, GeoArabia, Special Publication 2, Gulf Petrolik. Manama, Bahrain; 2001. 371 p

[6] Bellen RC, van Dunnington HV, Wetzel R, Morton DM. Lexique stratigraphique international. Asie, Iraq: International Geology Congress Commission de Stratigraphique, Fasco.10a. 1959;**3**:333

[7] Buday T. The regional geology of Iraq. Stratigraphy and Paleogeography. Dar Al-Kutub pub. University of Mosul, Iraq. 1980:**1**:445

[8] Jassim SZ, Goff, JC. Geology of Iraq, Published by Dolin, Prague and Moravian Museum, Brno. Printed in the Czech Republic. 2006. 341 p

[9] Van den-Kerkhof AM, Hein UF. Fluid inclusion petrography. In: Andersen T, Frezzotti ML, Burke EA, editors. Fluid Inclusions: Phase Relationships-methods-applications (Special Issue). Lithos. 2001;**55**(1-4):320

[10] Shingaly WS, Al-Juboury AI, Elias ME. The role of carbonate-host rocks on the genesis of lead-zinc deposits in Kurdistan Region, Northern Iraq. Journal of Tethys. 2015;**3**(1):31-47

[11] Bodnar RJ, Reynolds TJ, Kuehn CA. Fluid inclusion systematics in epithermal systems. In: Berger BR, Bethke PM, editors. Reviews in economic geology: geology and geochemistry of epithermal systems. Society of Economic Geologists. 1985;**2**:73-97

[12] Bodnar RJ, Vityk MO. Interpretation of microthermometric data for H_2O-NaCl fluid inclusions. In: De Vivo B, Frezzotti ML, editors. Fluid Inclusions in Minerals: Methods and Applications. Blacksburg: Virginia Technical Publication; 1994. pp. 117-130

[13] Roedder E. Fluid Inclusions, Mineralogical Society of America, Reviews in Mineralogy; 1984;**12**:644

[14] Leach DL, Sangster DF. Mississippi valley-type lead–zinc deposits. In: Kirkham RV, Sinclair WD, Thorpe RI Duke JM, editors. Mineral Deposit Modeling. Geological Association of Canada, Special Paper; 1993;**40**:289-314

[15] Paradis S, Hannigan P, Dewing K. Mississippi Valley-Type lead-zinc deposits. In: Goodfellow WD, editor. Mineral Deposits of Canada: A Synthesis of Major Deposit-types, District Metallogeny, the Evolution of Geological Provinces, and Exploration Methods. Geological Association of Canada, Mineral Deposits Division. 2007;**5**:185-203

[16] Leach DL, Sangster DF, Kelly KD, Large RR, Garven G, Allen CR, Gutzmer J, Walters S. Sediment-hosted lead–zinc deposits: A global perspective. Economic Geology. 2005;**100th** Anniversary Volume:561-607

[17] Leach DL, Taylor RD, Fey DL, Diehl SF, Saltus RW. A Deposit Model for Mississippi Valley-Type Lead-Zinc Ores, Chapter A of Mineral Deposit Models for Resource Assessment, USGS, Sci. Inv. Report 5070A. 2010. 52 p

[18] Taylor M, Kesler SE, Cloke PL, Kelly WC. Fluid inclusion evidence for fluid mixing, Mascot-Jefferson City zinc district, Tennessee. Economic Geology. 1983;**78**:1425-1439

[19] Gratz JF, Misra KC. Fluid inclusion study of the Gordonsville zinc deposit, central Tennessee. Economic Geology. 1987;**82**:1790-1804

[20] Hanor JS. The sedimentary genesis of hydrothermal fluids. In: Barnes HL, editor. Geochemistry of Hydrothennal ore Deposits. New York: Wiley Inter-science; 1979. pp. 137-142

[21] Schütfort EG. The Genesis of the San Vicente Lead Zinc Rhythmite Deposit, Peru—A Petrologic, Geochemical, and Sulfur Isotope Study [thesis]. USA: Oregon State University; 2001

[22] Ohmoto H, Rye RO. Isotopes of sulphur and carbon. In: Barnes HL, editor. Geochemistry of Hydrothermal Ore Deposits. New York: Wiley-Interscience; 1979. pp. 509-567

[23] Eldridge CS, Williams N, Walshe JL. Sulfur isotope variability in sediment-hosted massive sulfide deposits as determined using the ion microprobe SHRIMP: II. A study of the H.Y.C. deposit at McArthur river, Northern Territory, Australia. Economic Geology. 1993;**88**:1-26

[24] Sherlock RL, Roth T, Spooner ET, Bray CJ. Origin of the Eskay Creek precious metal-rich volcanogenic massive sulfide deposit: Fluid inclusion and stable isotope evidence. Economic Geology. 1999;**94**:803-824

[25] Roth T, Taylor BE. Sulfur isotope and textural zoning of pyrite in mudstone about the polymetallic Eskay Creek Deposit, Northwestern British Columbia, Canada. In: Volcanic Environments and Massive Sulfide Deposits International Conference; Hobart, Tasmania. 2000. pp. 16-19

[26] Al-Qaraghuli N, Lang H. Geochemical and mineralogical investigation of pyrite-sphalerite-galena deposit in Duri-Serguza area, Northern Iraq. Geological Society of Iraq. 1978;**11**:46-66

[27] He L, Song Y, Chen K, Hou Z, Yu F, Yang Z, Wei J, Li Z, Liu Y. Thrust-controlled, sediment-hosted, Himalayan Zn–Pb–Cu–Ag deposits in the Lanping foreland fold belt, eastern margin of Tibetan Plateau. Ore Geology Reviews. 2009;**36**:106-132

Chapter 4

Geology, Textural Study, Ore Genesis and Processing of the Tabuaço Tungsten Deposit (Northern Portugal)

Yann Foucaud, Bénédicte Lechenard,
Philippe Marion, Inna Filippova and Lev Filippov

Additional information is available at the end of the chapter

http://dx.doi.org/10.5772/intechopen.71674

Abstract

The Tabuaço tungsten deposit (Northern Portugal) is hosted in the Cambrian Douro Group metasediments, at the northern margin of the Beira-Tabuaço granitic complex. The hosting schisto-calcareous Lower Cambrian Bateiras Formation underwent a contact metamorphism induced by the intrusion of the granitic complex. The skarnification led to the crystallisation of scheelite ($CaWO_4$). Two different skarn facies are encountered: 'Lower Skarn' and 'Main Skarn' both corresponding to the exoskarn. The 'Main Skarn' is mainly composed of vesuvianite ($Ca_{10}Mg_2Al_4(SiO_4)_5(Si_2O_7)_2(OH)_4$), feldspars and fluorite. Zoisite, grossular, fluorapatite and scheelite are also present, as well as malayaite and cassiterite. Scheelite is disseminated and often occurs in association with fluorite, albite and vesuvianite. The 'Lower Skarn' level contains predominantly diopsidic pyroxene, quartz, zoisite, grossular and feldspars. Scheelite appears both laminated and disseminated, in association with fluorite and vesuvianite in minor proportions. The 'Main Skarn' is located in the Garnet, Pyroxene zone, while the 'Lower Skarn' corresponds to the Pyroxene, Garnet zone. Pyroxene has a global hedenbergite Hd_{40} composition. A major phase of albitisation destabilised fluorite, scheelite, vesuvianite, garnet and pyroxene. A late stage of chloritisation is associated with the exhumation. The processing of the Tabuaço ore has been adapted to the Ca-bearing rich paragenesis and to the scheelite mean size.

Keywords: skarn, scheelite, texture, paragenesis, processing

1. Introduction

The European Union has classified tungsten in 2010 as a critical raw material (CRM) [1]. This decision is based not only on the very high economic importance of tungsten in the European industry but also on the high supply risk to the member countries [2]. Indeed, in 2010, among the 16 world tungsten producers, China supplied 86% of the 72,100 tons of the global production [3].

This current situation results from the precipitous fall of tungsten price caused by Chinese dumping of low-operating-cost tungsten into global markets in the 1980s (**Figure 1**) [4]. Price remained low until 2005, preventing from any mining investment in other countries. Since then, China imposes quotas on its tungsten exports to preserve its own industry as a result of the growing worldwide domestic demand [3]. As a consequence, tungsten price has increased considerably for the last two decades, stimulating a gradual upturn of the exploration (**Figure 1**).

Tungsten is extracted from wolframite ((Fe, Mn, Mg)WO_4) and scheelite ($CaWO_4$). These two minerals can be found in Sn-W quartz veins, often associated with the Variscan orogenic cycle [5]. However, the major occurrence of scheelite is in skarn-type and skarnoid-type ores, related to a thermo-metamorphism of carbonated rocks. Historically, tungsten ores were predominantly processed by gravity techniques, often combined with magnetic separation, as W-bearing mineral density contrasts with gangue minerals [6]. Over the past decades, the explored tungsten ores have exhibited more and more fine-grained and in close association with gangue minerals, leading to the development of flotation as the main beneficiation route [7–9]. Nowadays, tungsten concentrates, assaying 20 to 65% WO_3, are mostly processed by hydrometallurgy and impurities such as phosphorus, silica and fluorine must be removed as much as possible in the preceding steps [10–12]. Mineral processing stage is consequently crucial in the elimination of the contaminant-bearing gangue minerals prior to hydrometallurgy. During beneficiation, W-bearing minerals are separated from the gangue minerals by physical processes, such as gravity and magnetic separation, and by physico-chemical processes as flotation. Performance of such separations requires a thorough knowledge of

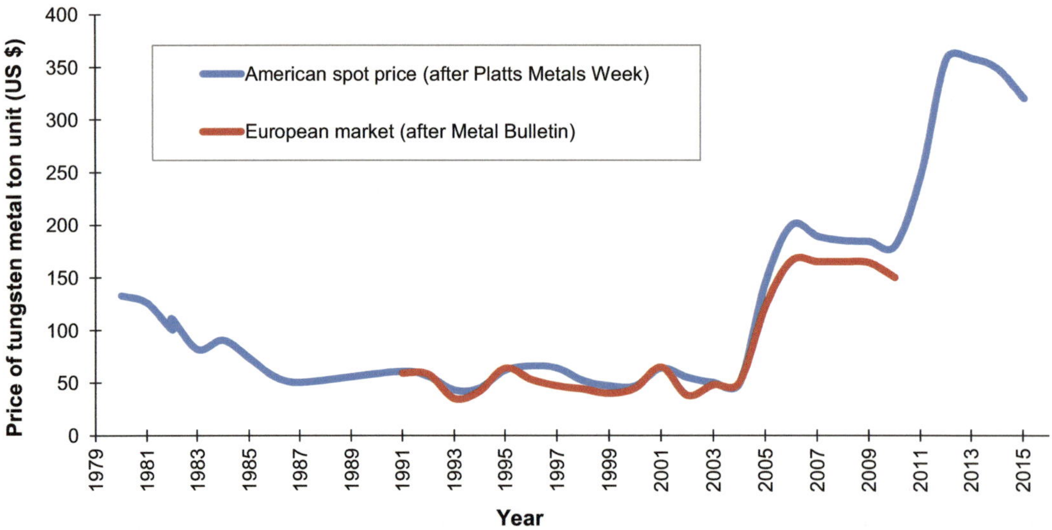

Figure 1. Evolution of tungsten metal ton unit price over the past decades.

the ore mineralogy including mineralogical and textural characterisation. Thus, the adoption of a systemic geometallurgical approach integrating studies from exploration to metallurgy ensures the best recovery of the valuable tungsten metal. This chapter presents mineralogical and textural studies performed on the Tabuaço tungsten skarn ore as a basis of the ore metallogeny. Results are used to design the optimum ore separation process.

2. Regional geology

2.1. Centro-Iberica zone

The Tabuaço deposit is located in the Central Iberian Zone (CIZ), one of the six zones composing the Iberian Massif (**Figure 2**), which is the part of the Variscan Belt in the Iberian Peninsula

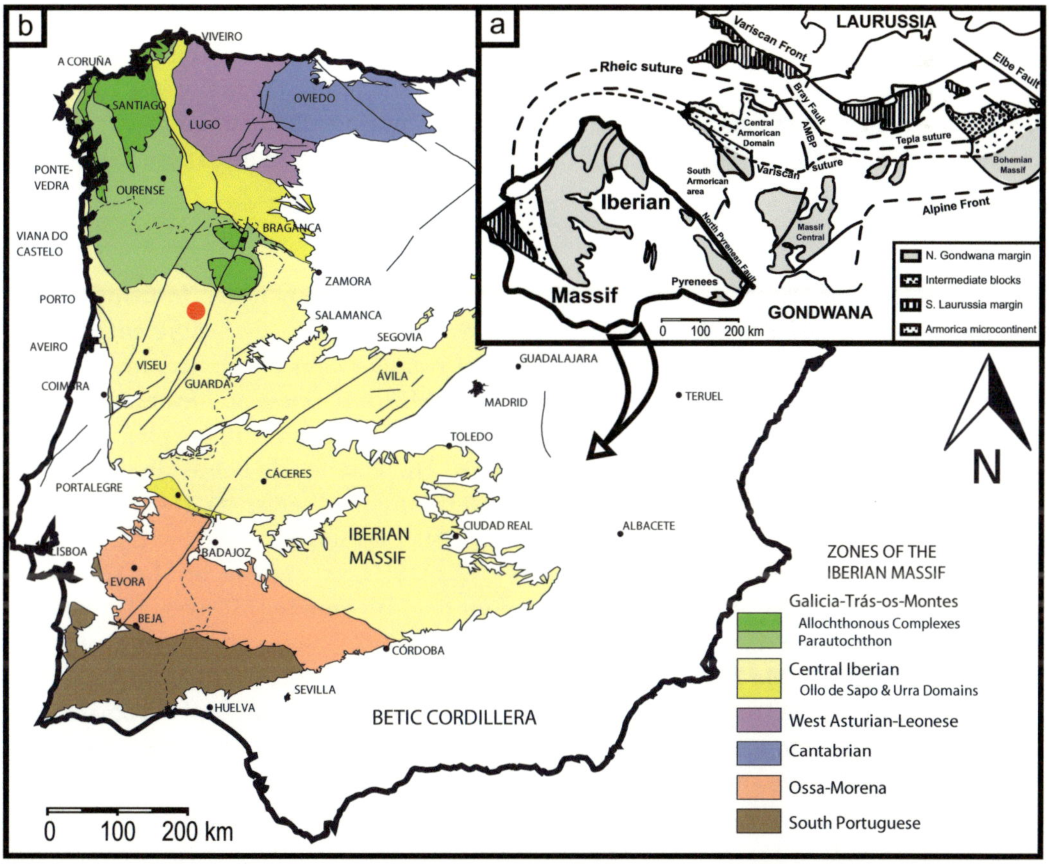

Figure 2. (a) Simplified tectonic sketch of the West European Variscan Belt showing the main occurrences of high-pressure rocks, modified after Faure et al. [13]. (b) Map showing the outcrops of the Variscan basement in the Iberian Peninsula and the zones of the Iberian Massif, modified after Martínez Catalán [14]. Red circle shows the location of the deposit.

[14, 15]. Details of the geodynamic evolution of the Variscan Belt have been extensively described [16–18]. The closing of the Iapetus and Rheic oceans during the Silurian-Devonian induced a collision between the two supracontinents Gondwana and Laurussia. The different zones were sutured to form a vast chain, covering all the Western Europe (**Figure 2**), called the Variscan Belt. The majority of the inner part of this chain, including the studied area, underwent a polyphased tectono-metamorphic evolution, going through at least three deformation stages and a regional green schist metamorphism [14].

2.2. Sedimentary formations

The Bateiras Formation is part of the Douro Group metasediments which are included in the Dúrico-Beirão Supergroup, also named 'Schist-Greywacke Complex' (SGC). The Douro Group is characterised by metapelitic rocks interlayered by metagraywacke and calcsilicate rocks [19]. The Bateiras Formation is constituted by a succession of metagraywackes/meta-conglomerates and black schists at the basis and by a succession of graphite-bearing black schists, black schists and calcareous layers [20, 21] going towards the youngest direction. The layers are centimetric to metric, for a global thickness of 800 ± 100 m [19]. This formation can be attributed to a terrigenous sedimentation at the basis, changing gradually to a basin sedimentation, occurred between 565 and 550 Ma [22, 23]. The whole Dúrico-Beirão Supergroup has then undergone the Variscan regional deformation and metamorphism described before.

2.3. Granitic intrusives

The intrusion of the Paredes da Beira-Tabuaço granite (PBT) leads to the formation of the Tabuaço W-skarn. The batholith extends over 200 km^2 and is zoned by three different medium-grained facies, known as Paredes da Beira-Tabuaço granite (outer zone), Sendim-Laboreira granite (middle zone) and Aricera granite (inner zone) [21, 24]. Some aplitic and pegmatitic bodies, closely associated with the granite, were reported by previous authors [25], associated with the crystallisation of the last magmatic liquids. The Paredes da Beira-Tabuaço granite stricto sensu contains quartz, two feldspars (Na-feldspar and K-feldspar), two micas (muscovite and biotite) and apatite [25]. The mineralogical association suggests that the granite is peraluminous as confirmed by a $1.08 < A/CNK < 1.32$ [25]. Low contents of MgO (≤ 0.06%), CaO (≤ 0.6%) and $(Fe_2O_3)_{tot}$ (≤ 1%) were reported for the PBT granite [25]. The PBT granite is part of Tabuaço granite massif, considered as tardi-D3 [25, 26], and yields K-Ar biotite and muscovite ages of 310 Ma [26]. Most recently, Rb-Sr. system is used on bulk on aplitic and pegmatitic bodies as well as on the Tabuaço granite yields with age of 311 ± 7 Ma [25, 27]. The Sr. and Nd isotopic compositions prove that these S-type granites are set up from a parental magma generated by the anataxis of the Grupo do Douro metasediments [25, 27].

3. Deposit geology

The Tabuaço tungsten deposit is located at the northern margin of the PBT granite and hosted in metamorphic rocks from the Bateiras Formation (**Figure 3**). It consists of two

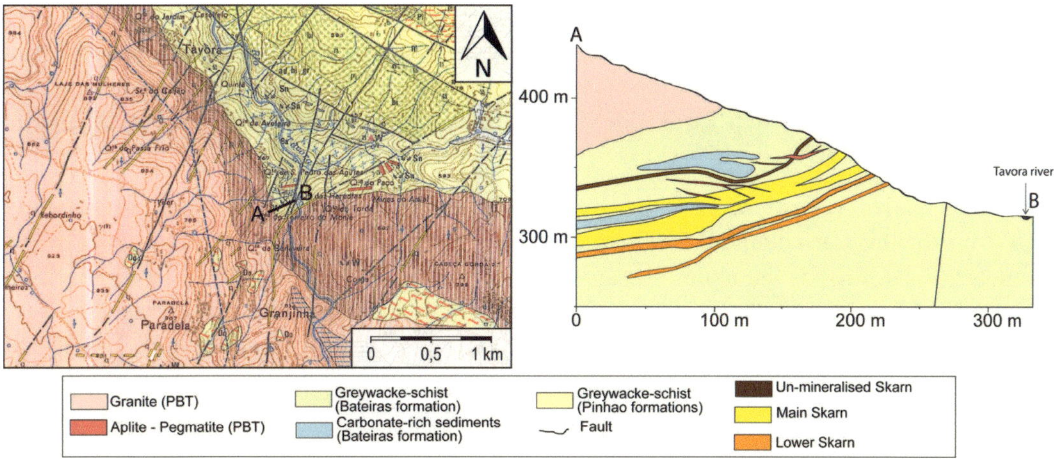

Figure 3. Left: Geological map of the studied area showing the Paredes da Beira-Tabuaço (PBT) granitic intrusion and the hosting Bateiras sedimentary formation (source: LNEG). Right: NE-SW cross section along the A-B line, based on Colt Resources drilling data and showing the main geological units of the Tabuaço deposit.

subdeposits referred as Aveleira and São Pedro das Águias. The study focusses on the latter, which represents 71% of the whole Tabuaço deposit. São Pedro das Águias deposit is composed of around 100 m of sedimentary rocks in which two mineralised units are included. Colt Resources reported in 2011 that a NI 43–101 indicated and inferred resource of 2 M.t. with a 0.56% WO_3 average grade, ranking Tabuaço as a middle-sized world-class tungsten deposit [28]. The two mineralised units, called 'Lower Skarn' and 'Main Skarn', are subparallel to the contact between the PBT granite and the Bateiras Formation, with a global strike-dip of N140–20°SW (**Figure 3**). The two horizons are completely different in terms of mineralogical association (presence/absence of pyroxene, abundance of garnet, etc.) and tungsten average grade (0.3% WO_3 for the Lower Skarn, 1% WO_3 for the Main Skarn), which can be related to their distance to the granite intrusion. The PBT granite intrusion occurred during the third tectono-metamorphic stage of the Variscan Belt, during which the global strike of shear zones, faults and fold axis was NW-SE [14]. Hence, the granite intrusion and the related skarn-hosted Tabuaço mineralisation are controlled by these structural deformations (**Figure 3**).

4. Sampling and analytical methods

Diamond drill programmes were performed by Colt Resources. Cores were used to characterise the deposit geology in terms of textures and mineralogy. Based on these analyses, a sampling campaign has been organised, and 330 kg of each of the two outcropping mineralised layers were sampled. On both the drill cores and the samples, thin polished sections were realised, scanned and analysed with, firstly, an optical microscope and then with a scanning electron microscope (SEM) Hitachi S-4800. This SEM utilises an electron beam accelerated from 500 V to 30 kV. The acceleration voltage used there was 15 kV.

Local chemical analyses on minerals were performed using a CAMECA SX100 Castaing electron microprobe. The acceleration voltage used there was 15 kV. Each analysis lasted 3 min and permitted to measure the composition of 15 elements in all the studied minerals, with a lateral resolution lower than 1 μm. Based on the results, structural formula of each mineral was then reconstituted.

Chemical analyses were carried out by inductively coupled plasma-mass spectrometer (ICP-MS) for the trace elements and by inductively coupled plasma-optical emission spectrometer (ICP-OES) for the major elements. Fluorine was analysed by direct ionometry and CO_2 by IR spectroscopy after combustion, with a C-S analyser. All these analyses were performed on representative pulverised samples at the Service d'Analyse des Roches et des Minéraux (SARM-CNRS, Nancy, France).

5. Results

5.1. Geochemistry of the mineralised layers

Geochemical analyses show that the two skarn units are significantly different in terms of chemistry, which partly explains the variety of mineral paragenesis (**Table 1**). The SiO_2 content within the calcareous rocks controlled the mineralogical assemblage resulting from the metasomatism event: silicates (including quartz) are predominant in the Lower Skarn, while the Main Skarn is enriched in calcium-bearing silicates and calcium-semi-soluble salts. In addition, the Lower Skarn has a higher Fe_2O_3 content and a lower CaO content than the Main Skarn (**Table 1**). Such differences in the chemistry between the skarn horizons must relate to the variation in chemistry of the protoliths since the PBT granite is too poor in Fe or Ca to induce such enrichments.

Notable W grade variation exists between the two skarn horizons (**Table 1**), the Main Skarn being more than eight times richer compared to the Lower Skarn. The differences in protolith chemistry and distance from the granite intrusion are two main controls of the W grade. As the Main Skarn is more proximal to the granite than the Lower Skarn

	SiO_2 (%)	Al_2O_3 (%)	Fe_2O_3 (%)	MgO (%)	CaO (%)	Na_2O (%)	K_2O (%)	P_2O_5 (%)	F (%)	CO_2 (%)	As (ppm)	W (%)	LOI (%)	Tot (%)
Main Skarn	37.81	17.45	2.39	1.66	31.56	1.32	0.72	0.73	5.6	0.18	18.8	1.7	5.09	98.98
Lower Skarn	47.25	17.14	6.06	2.31	21.38	1.86	0.94	0.33	1.47	0.16	180.7	0.23	2.17	100.16

Table 1. Chemical analyses showing major elements, some trace elements, F and CO_2 contents of the two mineralised layers.

(**Figure 3**), and presents higher CaO content, fluids were W-richer and scheelite precipitated preferentially.

Arsenic distribution within the skarns is also spatially zoned (**Table 1**). Very low grades are recorded in the Main Skarn, ten times lower than the ones in the Lower Skarn. Arsenic likely origins are from magmatic fluids associated with the granite intrusion, and mineralisation occurs in the vicinity or at the immediate proximity of a reduced layer.

5.2. Macroscopic scale study

The 'Main Skarn' unit presents massive, coarse-grained textures, strongly overprinting the almost lost sedimentary features during the metamorphism events (**Figure 4a**, **b** and **d**). Average grain size ranges between few micrometres to more than 1 cm. Some rock samples show weakly banded local features, which can be interpreted as a relic of the varying chemical composition between the layers of the Bateiras Formation or as a late albite vein-controlled deposition (**Figure 4b** and **c**).

In terms of mineralogy, sizeable amounts of honey-coloured fluorite are clearly visible macroscopically, associated with other minerals in pluri-millimetric patches (**Figure 4a**, **b**). In the samples, scheelite is easily identifiable under a short-wavelength UV-ray lamp, as it is a fluorescent mineral (**Figure 4**). Scheelite is mostly finely disseminated, forming micrometre- to centimetre-scaled crystal aggregates (**Figure 4a**, **b** and **d**). These aggregates are probably composed of an association between scheelite and other minerals, the accurate spotting of scheelite being difficult with UV-ray lamp due to the diffusion of the fluorescence light.

The 'Lower Skarn', located in the distal part of the skarn deposit, shows quite well-preserved sedimentary features. Macroscopic banded textures (**Figure 4e–h**) are interpreted as a result of the bimetasomatism phenomenon, developed between centimetric sedimentary layers with varying chemical compositions. The preservation of the initial macroscopic structure is directly linked to the temperature gradient from the intrusion heat source and then the degree of the bimetasomatism. This latter has been nearly complete in the Main Skarn but only partial in the Lower Skarn. However, it has been powerful enough to permit the formation of calcic silicates in all the layers (**Figure 4e** and **h**).

In the Lower Skarn, macroscopic identification is much more difficult because of the very fine-grained texture. However, centimetric patches of vesuvianite, a common Ca-bearing hydrated silicate in skarns, are visible associated with pinkish K-feldspars and green pyroxene (**Figure 4g**). Even if zoisite, a calcic epidote-group mineral, is stable in all the mineralogical association, some light-green layers are notably enriched (**Figure 4e**). As well, pyroxene-rich and vesuvianite-rich zones are alternating in some samples (**Figure 4f** and **h**). Green zones contain pluri-millimetric sulphide crystals, being mainly arsenopyrite and pyrrhotite (**Figure 4d** and **f**). Such zones are rare in the skarns and represent some reduced levels, with high sulphur content, weakly affected by the metasomatic and metamorphic processes. An

Figure 4. Photographs of hand samples showing major mineralogical assemblages and textures of the Main Skarn (a to d) and the Lower Skarn (e to h) under natural light associated with a short-wavelength UV-ray lamp. (a) Disseminated scheelite with large patch of honey-coloured fluorite. (b) Laminated scheelite close to albite and fluorite. (c) Banded texture showing alternating zoisite-rich zones and albite veins. (d) Penetrative texture of albite into vesuvianite + scheelite + fluorite assembly. (e) Inherited banded texture with laminated scheelite along the sedimentary layer boundaries. (f) Sulphide-rich green zone presenting a succession of vesuvianite-rich and pyroxene-rich zones. (g) Progressive change from a pyroxene-rich zone to a K-feldspar + vesuvianite zone, associated with albite. (h) Scheelite occurring both as disseminated and laminated and associated with a pyroxene-rich zone and vesuvianite patches. Note the oxidisation front, separating the sulphide-rich zone (right) from the sulphide-depleted zone (left). Sch, scheelite; Ves, vesuvianite; Zo, zoisite; F, fluorite; Alb, albite; As, arsenopyrite; Po, pyrrhotite; Ox, oxides; Px, pyroxene; K-Fd, K-feldspars.

oxidation front clearly separates reduced sulphide-rich zones and oxidised sulphide-depleted zones (**Figure 4h**). In the oxidised zone, vesuvianite is clearly identified macroscopically, forming large areas of dark prisms.

In terms of W mineralisation, in the Lower Skarn, poorer than the Main Skarn, only rare occurrences of disseminated scheelite can be observed macroscopically (**Figure 4h**). Noteworthy, the two mineralised skarns (Lower and Main) also present laminated scheelite which seems controlled by the chemical composition of the hosting sedimentary layers (**Figure 4b** and **e**).

Albite is largely present in the two mineralised horizons, occurring in veins (**Figure 4b** and **c**) or associated with other minerals such as vesuvianite (**Figure 4g** and **d**). It seems that albite veins are penetrative into the skarns. A correlation exists macroscopically between the presence of albite and the absence of scheelite (**Figure 4d** and **e**).

5.3. Visible light and electron microscope scale study

Thin sections were observed under optical microscope and under scanning electron microscope. While scheelite appears laminated in hand samples, microscopically, all the observed scheelite is finely disseminated and forms anhedral crystals with sizes of between 5 and 300 μm for the Main Skarn (**Figure 5a** and **b**) and below 150 μm for the Lower Skarn. These values are considered as the liberation mesh required for an efficient ore separation process.

Molybdenum content was measured in the scheelite using the Castaing electron microprobe. Regardless of the Skarn horizon type, very low grades were found, 0 to 0.12% MoO_3 with an average of 0.029% MoO_3. The scheelite is zoned, with grain cores richer in MoO_3 than the rims.

Scheelite grains are closely associated with vesuvianite (**Figure 5a**; **Figure 7a**), which forms large areas, above all in the Main Skarn (**Figure 5a**), as the main calcium-bearing silicates of the global mineralogical association (Main and Lower Skarns). Structural formula shows no variation between the Lower Skarn and the Main Skarn, being $(Ca_{9.36}Mg_{0.41})(Al_{1.13}Fe_{0.80}Mg_{0.07})Al_4[(Si_{8.83}Al_{0.09})O_{32.39}F_{1.61}]$ with traces of strontium, manganese and phosphorus. Hence, fluoro-vesuvianite is stable in the whole skarn body, regardless of the temperature or the chemistry of the protoliths.

Close association between vesuvianite and euhedral garnets (**Figure 5c**) proves their synchronous formation. Besides, the fact that their chemical analyses are similar indicates that temperature rather than fluid chemical properties controls the precipitation of one mineral over the other one. Microprobe analyses give an average structural formula of $(Ca_{2.99}Sr_{0.01})(Al_{1.75}Fe_{0.25})[(Si_{2.87}Al_{0.1})O_{11.6}F_{0.4}]$ for all the analysed garnets, which are individually represented in a ternary diagram (**Figure 6**, left). Analysed garnets are from the Main Skarn horizon only, as very few garnets are present in the Lower Skarn. In summary, garnets are $Gr_{87}And_{13}$, proving the predominance of Al^{3+} and Ca^{2+} on Fe^{3+} and Fe^{2+}, respectively.

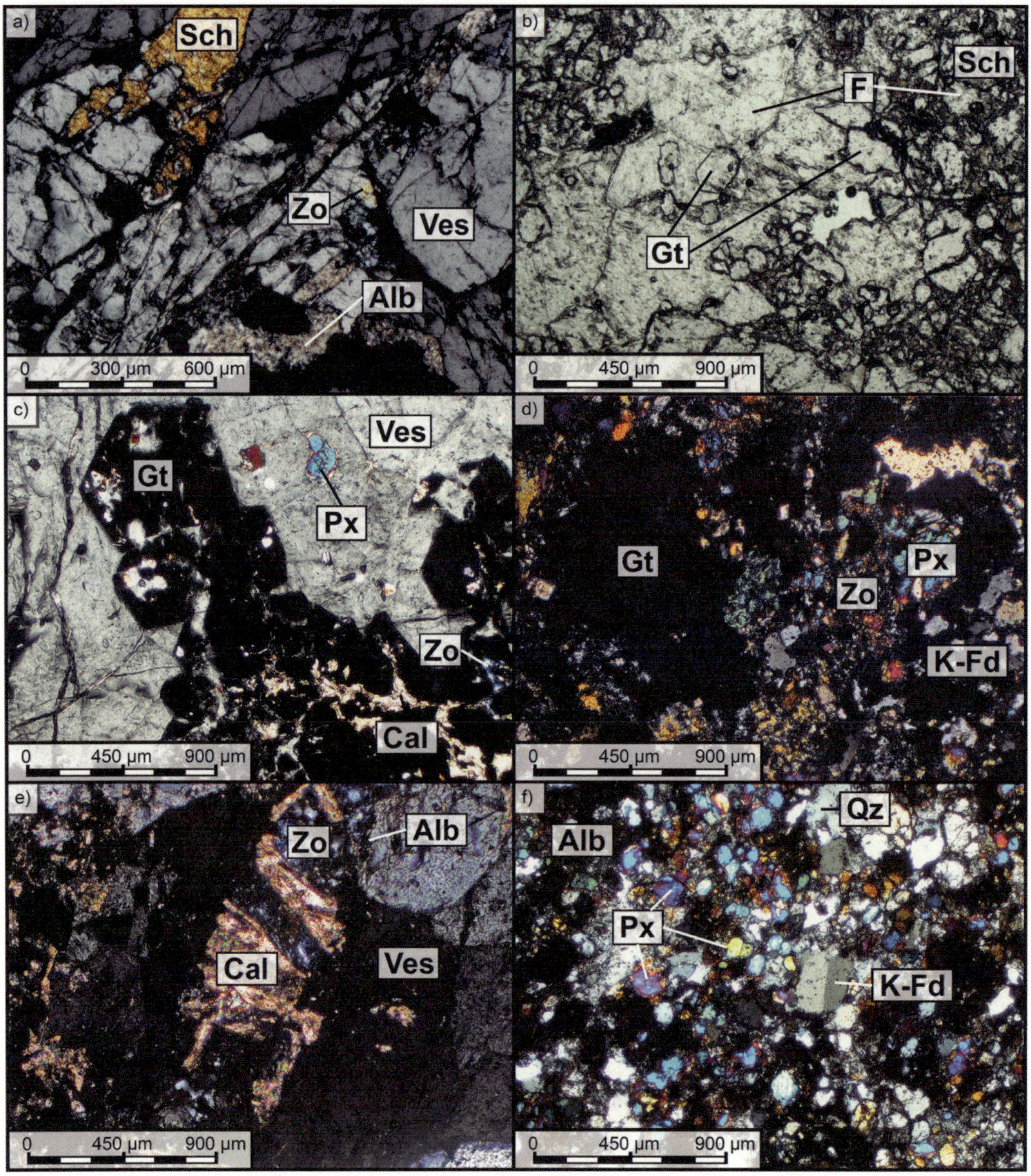

Figure 5. Photographs of thin sections observed with optical microscope in cross polarised light (a, c to f) and in polarised light (b) showing different microscopic textures and mineral assemblage in the Main Skarn (a to c) and in the Lower Skarn (d to f). (a) Vesuvianite area containing both scheelite and zoisite with a kaolinised albite alteration. (b) A typical association between scheelite, garnet and fluorite. (c) Synchronous formation of euhedral garnets and vesuvianite, containing a clinopyroxene, with a tardive setup of calcite + epidote. (d) Poikilitic texture of coexisting clinopyroxene and garnet, finely associated with K-feldspars and zoisite. (e) Vesuvianite crystal altered into albite with calcite + epidote overprint. (f) Typical mineral association of the Lower Skarn, composed of pyroxene, feldspars, zoisite and minor quartz. Same legend as **Figure 4** with Qz, quartz; Cal, calcite; Gt, garnet.

Very rare crystals of clinopyroxene are present in the Main Skarn level, often included in vesuvianite grains (**Figure 5c**). In the Lower Skarn, clinopyroxene is predominant over garnet. However, some poikilitic textures of garnets surrounding pyroxene are found (**Figure 5d**), proving the coexistence of both in the Lower Skarn, where the pyroxene is closely associated with quartz, K-feldspars and zoisite (**Figure 5f**). The centimetric layers observed macroscopically have been described as alternating pyroxene-rich levels, inducing the green colour, and quartz/K-feldspars/vesuvianite-rich levels (**Figure 4f**, **g** and **h**). Optical factors indicate that the pyroxene has an intermediate composition between hedenbergite (Fe) and diopside (Mg) poles of calcic clinopyroxene continuous series. This observation is confirmed by electron microprobe analyses, which permitted to calculate a global average formula of $Ca_{1.01}(Fe_{0.40}Mg_{0.59}Mn_{0.01})[(Si_{1.98}Al_{0.01})O_6]$ for the Lower Skarn, as no pyroxene of the Main Skarn could be analysed. The pyroxene is then $Hd_{40}Di_{59}Jo_1$.

Zoisite, clearly identified microscopically (**Figure 5a**, **c**, **d** and **e**; **Figure 7c** and **e**), is frequently zoned, probably admitting significant variations in the chemistry. Electron microprobe showed that zoisite has a significant part of the aluminium substituted by Fe^{2+} and Mg^{2+} (in smaller proportions) in the octahedral sites. The calculated structural formula, $Ca_{2.02}(Al_{2.59}Fe_{0.37}Mg_{0.01})[(Si_{2.99}Al_{0.01})O_{10.38}(OH)_{0.62}]$, indicates that it is a Fe-rich zoisite, called after Fe-zoisite. The Fe-zoisite is present in both Main and Lower Skarns and seems stable in the mineralogical association. Sizeable amounts of Fe-zoisite are associated with calcite in micro-veins (**Figure 5e**), postponing the albite previously described. These veins are cross-cutting the whole samples, visible macroscopically on the thin sections.

Another major mineral of the paragenesis is fluorite, which forms centimetric patches, often in association with vesuvianite and scheelite in the Main Skarn (**Figure 5b**; **Figure 7a**, **b** and **e**). Fluorite is also closely associated with pyroxene in the Lower Skarn while less abundant. Microprobe analysis shows that vesuvianite and apatite bear fluorine, in substitution of the hydroxyl group. They are then fluoro-vesuvianite and fluorapatite, respectively. Garnet also bears fluorine, in substitution of oxygen.

A late albitisation stage is responsible for the destabilisation of the metasomatism-related mineral association. This alteration is promoted by cross-cutting micro-veins of albite (**Figure 5a** and **e**; **Figure 7b**). Most of the albite is developed at the selvedge of the grains

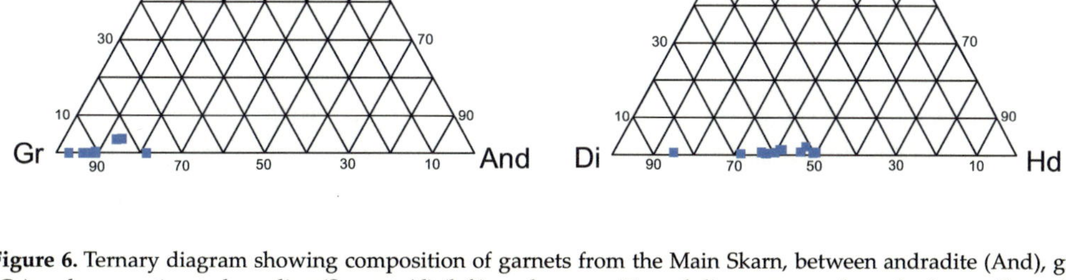

Figure 6. Ternary diagram showing composition of garnets from the Main Skarn, between andradite (And), grossular (Gr) and spessartine + almandine (Spess + Al) (left), and composition of clinopyroxene from the Lower Skarn, between hedenbergite (Hd), diopside (Di) and Johannsenite (Jo) (right).

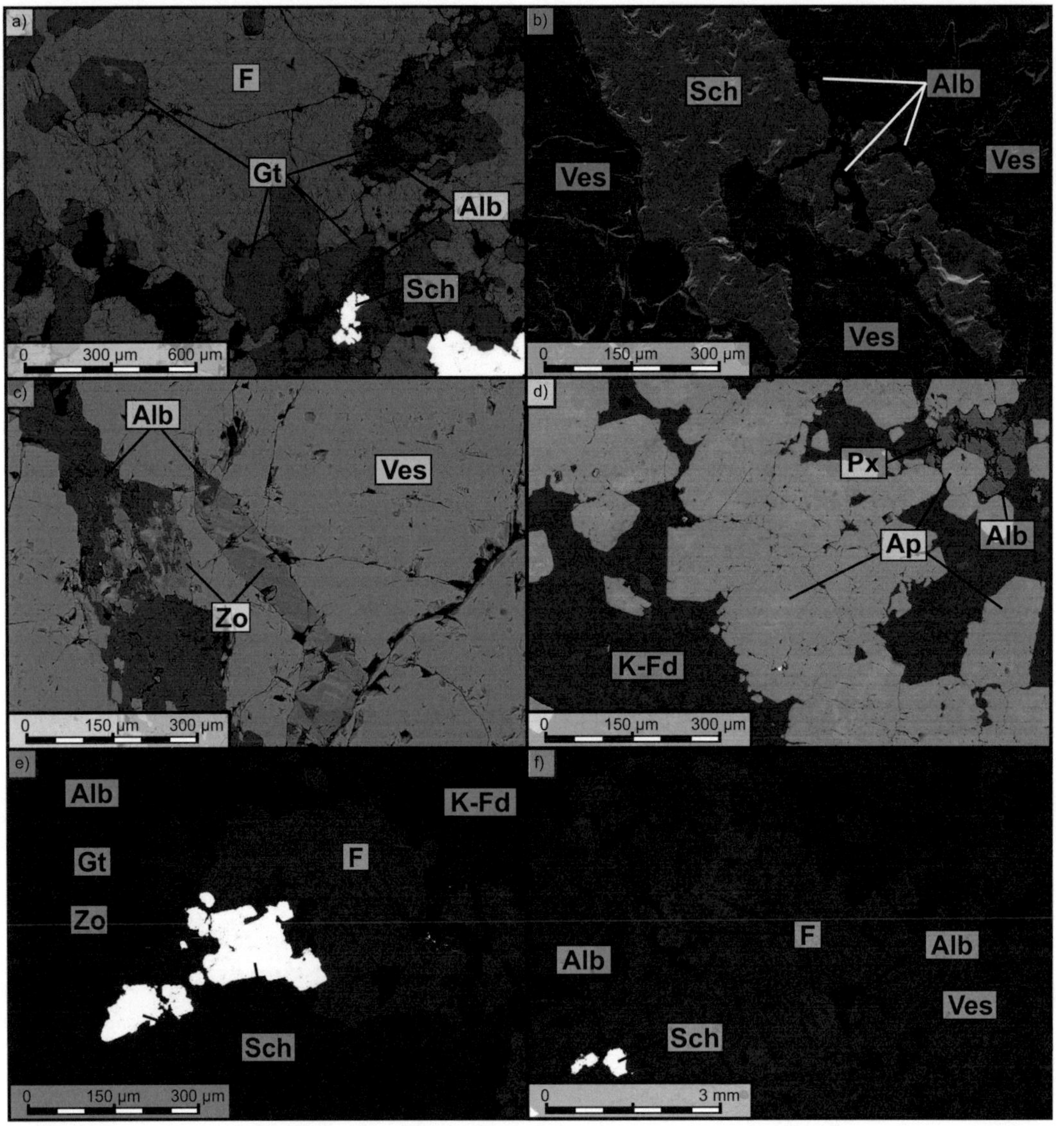

Figure 7. SEM images of Main Skarn thin sections, in back-scattered electrons imaging (a, c to f) and in secondary electron imaging (b), showing different textures on the ore. (a) Association of scheelite, garnet, fluorite and vesuvianite, the garnet being unsettled by albite. (b) Association of scheelite and vesuvianite, with an albite micro-vein unsettling vesuvianite at the grain joint. (c) Zonation in a zoisite, which is being unsettled by albite. (d) Apatite with K-feldspars and pyroxene, being consumed by albite. (e) Close association between scheelite, fluorite, garnet and zoisite, surrounded by albite. (f) Scheelite and fluorite being unsettled by albite. (Same legend as **Figures 4** and **5**).

(**Figure 7c** and **d**). The shape of the original crystal is preserved, forming phantoms of primary minerals (**Figure 7a**, **c** and **f**). Destabilisation textures, such as very fine intergrowth of albite along the 90° cleavages of fluorite at the beginning which generalises after (**Figure 7f**), can be observed for most of the minerals: garnet (**Figure 7a**), vesuvianite (**Figure 7b** and **f**), zoisite and even scheelite (**Figure 7f**).

6. Discussion

6.1. General discussion

Microprobe analyses indicate that scheelite is zoned in terms of Mo-content. This phenomenon is explained by the low sulphur content in the hosting sedimentary rocks, as no molybdenite was found. Molybdenum, in very low amount, originates from magmatic fluids and is incorporated into the scheelite crystals, contributing to the continuous solid solution between scheelite ($CaWO_4$) and powellite ($CaMoO_4$). The low content of molybdenum in scheelite suggests a high quality of this W deposit, as molybdenum represents a critical penalising element for tungsten metal properties. Some authors reported that Mo-poor scheelite is associated with oxidised skarns [29]. More recent studies proved that Mo-poor scheelite can as well form in reduced skarns [30–32]. Based on the later classifications [31], the Tabuaço skarn is a reduced skarn. Indeed, it presents Mo-poor scheelite which is related to a low dioxygen fugacity [30], high pyrrhotite:pyrite ratios and grossular-rich garnet (Gr_{87}). The andradite content is very low in the garnet, for which the Al^{3+}/Fe^{3+} ratio is around 7.4. Moreover, all the other Fe-bearing minerals are all Fe^{2+}-bearing minerals (vesuvianite, pyroxene and epidote). All these elements are suggesting that fluids going from the granitic intrusion to the hosting sedimentary rocks are reduced, with low dioxygen fugacity and then low Fe^{3+}/Fe^{2+} ratio.

Scheelite crystallisation is controlled by pH and Ca^{2+} activity in the hosting rocks, as the tungsten is transported by acidic magmatic fluids in the form of tungstic acid [30]. A brutal increase of the pH or of the Ca^{2+} concentration in the fluids, as it would happen when cross-cutting a carbonate-rich horizon, can lead to the precipitation of scheelite [30, 33]. Hence, laminated scheelite would have primary precipitated at the contact of carbonate-rich layers of the sediment. A fluid grade of 100 ppm of W is reported to be enough to cause scheelite precipitation at the vicinity of a carbonated layer [33]. Disseminated scheelite, sparser, likely results from a remobilisation of the laminated scheelite, postponing bimetasomatism reactions which led to the loss of the initial heterogeneity of the sedimentary hosting rocks.

The two mineralised horizons correspond to carbonate-rich layers alternating with thin levels of silicate-rich sediments. This succession allowed the crystallisation of calcium-bearing silicates by a bimodal exchange between the two categories of sediments (bimetasomatism). However, the Lower Skarn protolith was probably more silicate-rich than the Main Skarn protolith. The dominance of calcium in Main Skarn induced the formation of sizeable amounts of semi-soluble salts, as scheelite, fluorite and apatite. These two latter crystallised as a result of an increase of Ca^{2+} and even more importantly F^- and PO_4^{3-} activities in the fluids. These volatile-enriched fluids correspond to end-magmatic crystallisation fluids and explain the substitution of hydroxyl group or of oxygen by fluorine in apatite, vesuvianite and garnet [30, 32]. The fluorine origin is probably sedimentary, as the granite is a peraluminous S-type granite, formed by anatexis of the Bateiras Formation. Some Mo-poor W-F skarns, derived from S-type granite intrusions, were described by authors [34]. The composition of fluids is generally intermediate between oxidised and reduced, and the skarn cannot be really classified [32].

Tin-bearing minerals are also present in minor amounts. It consists of cassiterite and malayaite, two calcium/tin-bearing silicates which presence has been several times reported by authors in a similar context [29, 30, 35].

Furthermore, the mineralisation is closely linked to the setup of the skarn induced by the intrusion of a S-type peraluminous tardi-D_3 of two mica granites (PBT granite). In the late stage of crystallisation, fractionated fluids were enriched in some elements (W, F, P). Their slow diffusion into the hosting rocks, from the granite to the metamorphosed sediments, was driven mainly by the temperature gradient and variations in chemical potential [30, 36]. As the hosting sediments were composed of silicate-rich and carbonate-rich centimetric to metric layers, very associated, each brutal change induced a rebalancing of the fluids with the rock (**Figure 9**). Scheelite, fluorite and apatite precipitated when both the Ca^{2+} activity and the pH increased suddenly, at the contact with the carbonated layers (**Figure 9**). Because of its proximal location to the granite, the temperature is higher in the Main Skarn than in the Lower Skarn, inducing a faster and nearly complete homogenisation of the initial banded textures. Difference in temperature also explains the dominance of the garnet in the Main Skarn and the dominance of the pyroxene in the Lower Skarn, consolidated by chemical contrast. Indeed, the crystallisation of pyroxene requires more SiO_2 compared to the garnet, and bulk analyses showed that the Lower Skarn is SiO_2-richer (**Table 1**). In addition, the graphite content seems

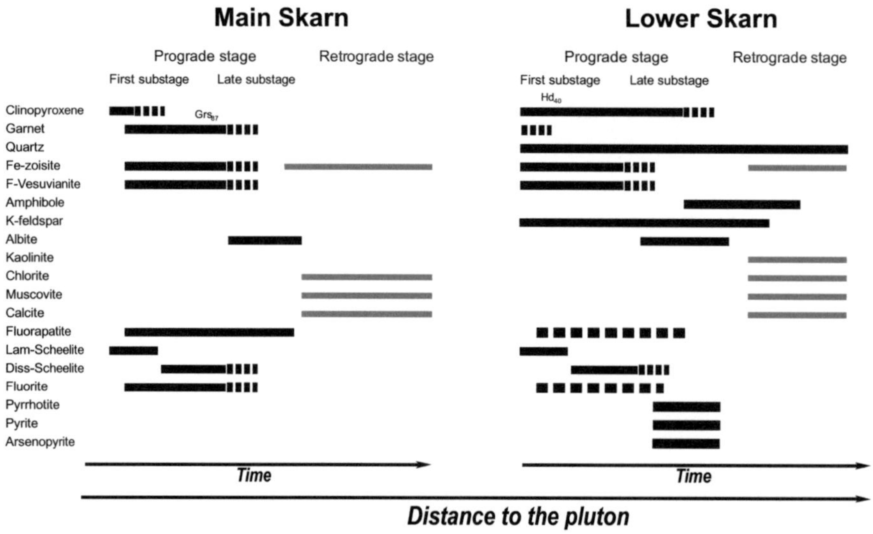

Figure 8. Paragenetic sequence of the Tabuaço scheelite skarn, showing the evolution of the mineral associations versus time and versus distance to the pluton (i.e. for the two skarn horizons).

to play a role in the variation of skarn composition [30, 37] as a graphite-rich layer will lead to the formation of Pyr > Gar and pyrrhotite > > pyrite zone as observed in the Lower Skarn. Finally, the Tabuaço skarn is spatially zoned (**Figure 9**) with a Gar > Pyr zone transitioning to a Pyr < Gar zone. The latter zone contains sizeable amounts of sulphides, mainly pyrrhotite and arsenopyrite, which reflects the high sulphur and probably graphite content of the hosting sediments (**Figure 9**). It could correspond to the graphite-bearing shales, occurring in the Bateiras Formation.

The skarn zonation has been intensively described by authors [30, 32, 35, 38] and could be summarised as follows, from the pluton to the metamorphic marbles: **Quartz ± Garnet (inferred) → Garnet > Pyroxene (Main Skarn) → Pyroxene > Garnet (Lower Skarn) → Pyrrhotite zone → Wollastonite ± Quartz ± Calcite (inferred) → metamorphic marbles (Figure 9).**

No evidence of wollastonite was found in the mineralogical observations. The presence of high content in aluminium, due not only to the peraluminous nature of the granite but also to the chemistry of the sediments, induced the formation of aluminium-bearing calc-silicates as vesuvianite, garnet and epidote, more favourably than the formation of wollastonite. But,

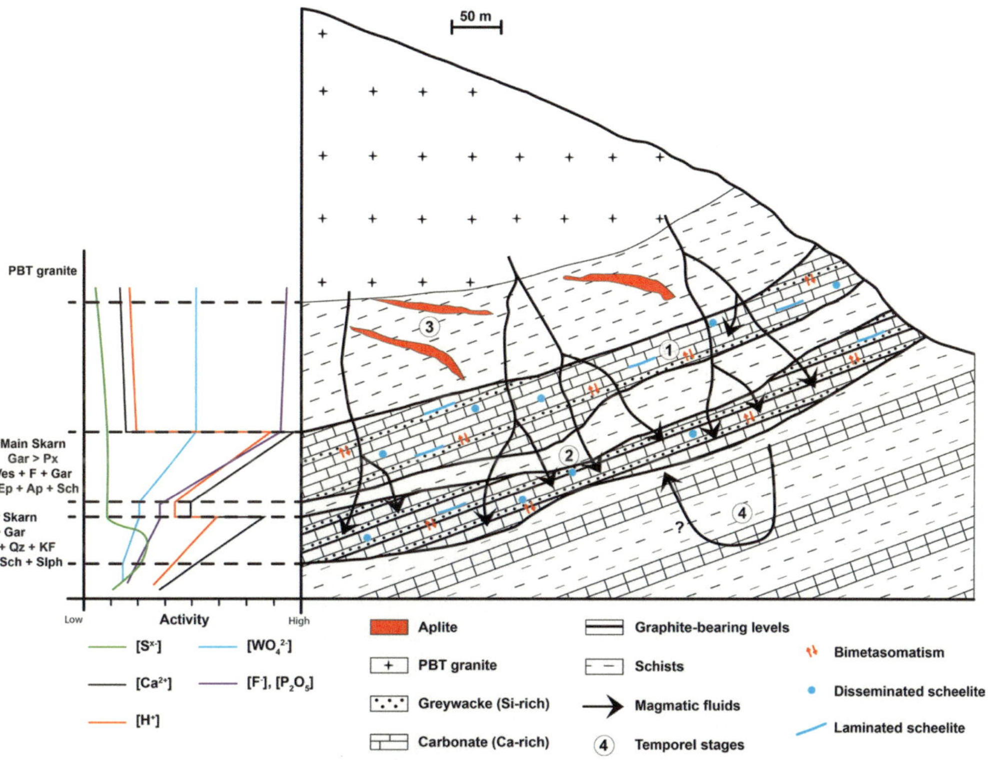

Figure 9. Schematic cross section of the Tabuaço deposit showing the spatial zonation and the temporal stages of mineral association. (See the text for further explanations.) Gar, Garnet; Px, pyroxene; Ves, vesuvianite; F, fluorite; Ep, epidote; Ap, apatite; Sch, scheelite; Qz, quartz; KF, K-feldspars; Slph, sulphides.

distal to the pyrrhotite-rich zone, described by the authors [30, 35], a wollastonite-rich zone could exist, even if it may be absent in several world skarns [39].

Within this zonation, vesuvianite is present and stable everywhere in the Gar > Pyr and in the Pyr > Gar zones, as well as zoisite (**Figure 9**). These two minerals are primary, based on the mineralogical observations, and associated with other primary minerals as fluorite, scheelite, apatite and garnet/pyroxene. Authors reported that zoisite can be primary, as a replacement of anorthite feldspar often described but absent in the Tabuaço mineral association [30]. As well, fluorite and apatite are stable in both the Main and Lower Skarns, their different contents being explained by the depletion of the F and P contents in the fluids reaching the distal part of the skarn (**Figure 9**). The same phenomenon occurred with the W content (**Figure 9**).

Authors reported that the formation of garnet and pyroxene occurs at least for 300°C [30, 40], while the zoisite epidote does not form under 445°C [32]. Pyroxene is probably formed first and unsettled in the skarn evolution by the garnet, as poikilitic textures are observed in this ore and have been observed for other ores [32]. Pyroxene is transformed into garnet when the fluids become SiO_2-richer and, above all, saturated [30, 41], occurring in the early prograde evolution.

The late prograde stage is characterised by the circulation of Na-rich fluids, which corresponds to the last fluids, enriched in incompatible elements by the fractionated crystallisation (**Figure 9**). These fluids come probably with the important intrusion of aplite veins at the immediate vicinity of the granite. Na-rich fluids started to unsettle the primary mineral association, transforming garnets, pyroxene, fluorite, vesuvianite and even scheelite into albite. However, this phenomenon is only partial in the skarn.

Finally, the retrograde stage is represented by an inverse fluid circulation, introduced few decades ago [29]. In the Tabuaço skarn, this late stage is characterised by an epidote + calcite association, mainly formed in the form of cross-cutting veins, postponing all the primary minerals. Some white micas (muscovite) and chlorite-group minerals have been observed, corresponding to the retrograde late stage. Supergene alteration induced the kaolinisation of the majority of albite formed in the skarn.

7. Conclusion: integrated mineralogy into a geometallurgical approach

A detailed mineralogical study of the skarn-hosted Tabuaço deposit was performed on rock sampled from two different skarn horizons. Mineralogical assemblages and textural observations are key elements to integrate into the beneficiation process design.

To begin with, consideration of the liberation mesh is primordial for the optimisation of the grinding steps. Liberation mesh was evaluated by optical microscopy under short-wavelength UV lamp and confirmed by geochemical analysis of the different size fraction. Liberation was estimated at an average size of 150 μm for both units, while it is recognised that the Lower Skarn unit has slightly smaller scheelite grains averaging around 120 μm.

Then, the present study enabled to characterise mineralogical associations and crystallisation/alteration textures. In this regard, a feature of interest is the close association of the valuable scheelite with vesuvianite which is one of the major mineral phases. Metasomatism reactions led to the crystallisation of various calcium-bearing gangue silicates such as vesuvianite, garnet, pyroxene and epidote for the major ones. Thus, the main challenge in recovering scheelite versus calcium-bearing minerals via flotation processes is their selective separation due to the similarities in terms of surface properties, electrokinetics and stability/solubility in aqueous solutions. Conventional silicate depressants, for example, sodium silicate, exist, but the issue can only be resolved if the Ca^{2+} site-specific collector selectivity is high. Because of their different spatial relationship to the main granite intrusion along with the variation in protoliths lithology and chemical compositions, the two skarn horizons studied present distinct mineralogical assemblages. Thus, flotation reagents in terms of collector and depressant need to be tailored to the ore feed composition. No matter how successful the recovery of scheelite is, the concentrate grade can only be maximised if the grade of gangue minerals is reduced. For instance, fluorite is a major phase in the Main Skarn unit, and special effort should be taken to minimise its grade in the final concentrate.

Crystallisation textures are also essential to take into account in seeking and improving the concentrate grade. In particular, mineral intergrowth, poikilitic texture, micro-veins and secondary overprint are typical features of mixed particles occurring even if the liberation mesh is reached. Such mixed particles are likely to be harder to float, and in the eventuality that they are floated, the concentrate grade will be diluted by impurities.

Moreover, a pervasive albitisation of the skarn horizons, principally the Main Skarn, affects the whole prograde paragenesis, including scheelite. The overprint of albite onto scheelite grains inhibits the collector action, reducing significantly the concentrate grade and the overall recovery. In addition, albite has been subsequently altered into kaolinite and other clay minerals. These fragile minerals are concentrated in the very fine fractions (−10 μm) generated by the comminution steps; hence, it is crucial to perform a desliming step in order to remove this phase.

The optimisation of the flotation process should also consider the downstream treatments such as hydrometallurgy processes. In this regard, the concentration of penalising elements such as phosphorous should be limited. Fluorapatite, occurring predominantly in the Main Skarn horizon, requires the use of specific depressants to inhibit the action of Ca^{2+} site-specific collector. The Lower Skarn horizon is less challenging in terms of gangue mineral depression given the dominance of easily depressed silicates such as quartz, K-feldspar or pyroxene and the rare abundance of fluorapatite, fluorite or garnet.

Acknowledgements

The research leading to these results has received funding from the European Union's Horizon 2020 research and innovation programme under grant agreement No. 641650 for the

FAME project. This chapter reflects only the authors' view, exempting the commission from any liability. The authors would like to thank Gaëtan Fillon, Gabriel Crumiere and Frédéric Diot for their active collaboration in this work.

Author details

Yann Foucaud, Bénédicte Lechenard, Philippe Marion, Inna Filippova and Lev Filippov*

*Address all correspondence to: lev.filippov@univ-lorraine.fr

Université de Lorraine, GeoRessources, UMR 7359 UL-CNRS, Vandoeuvre-lès-Nancy, France

References

[1] European Commission. Critical Raw Materials for the EU: Report of the Ad-Hoc Working Group on Defining Critical Raw Materials. 2010. http://ec.europa.eu/enterprise/policies/rawmaterials/documents/index_en.htm

[2] Dewulf J, Blengini GA, Pennington D, Nuss P, Nassar NT. Criticality on the international scene: Quo vadis? Resources Policy. 2016;**50**(2016):169-176

[3] Audion AS, Labbé JF. avec la collaboration extérieure de la Compagnie Européenne d'Intelligence Stratégique (CEIS), 2012. Panorama mondial 2011 du marché du tungstène. Rapport Public. BRGM/RP-61341-FR, 108 p., 29 fig., 16 tabl. In French

[4] Lassner E, Schubert WD. Tungsten: Properties, Chemistry, Technology of the Element, Alloys and Chemical Compounds. 1999

[5] Jébrak M, Marcoux E. Geology of Mineral Resources. 2015

[6] Burt R. The role of gravity concentration in modern processing plants. Minerals Engineering. 1999;**12**(11):1291-1300

[7] Agar GE. Scheelite Flotation Process. Patent specification 4488959, USA, 17 p. 1982

[8] Atademir MR, Kitchener JA, Shergold HL. The surface chemistry and flotation of Scheelite. II. Flotation "collectors". International Journal of Mineral Processing. 1981;**8**:9-16

[9] Hiçyìlmaz C, Atalay Ü, Özbayoglu G. Technical note. Selective flotation of scheelite using amines. Minerals Engineering. 1993;**6**(3):313-320

[10] Pastor H. Métallurgie et Recyclage Du tungstène. 2000. Procédés. Dossier – Techniques de l'Ingénieur

[11] Yang Y, Xie B, Wang R, Xu S, Wang J, Xu Z. Extraction and separation of tungsten from acidic high-phosphorus solution. Hydrometallurgy. 2016;**164**(2016):97-102

[12] Zhang G, Guan W, Xiao L, Zhang Q. A novel process for tungsten hydrometallurgy based on direct solvent extraction in alkaline medium. Hydrometallurgy. 2016; **165**(2016):233-237

[13] Faure M, Cocherie A, Gaché J, Esnault C, Guerrot C, Rossi P, Wei L, Qiuli L. Middle carboniferous intracontinental subduction in the outer zone of the Variscan Belt (Montagne noire axial zone, French massif central): Multimethod geochronological approach of polyphase metamorphism. Geological Society, London, Special Publications. 2014;**2014**(405):239-311

[14] Martínez Catalán JR. The central Iberian arc, an orocline centered in the Iberian massif and some implications for the Variscan belt. International Journal of Earth Science (Geol Rundsch). 2012;**101**:1299-1314

[15] Julivert M, Fontboté J-M, Ribeiro A, Conde L. Mapa Tectónico de la Península Ibérica y Baleares E. 1:1,000,000. 1972. Inst. Geol. Min. España, Madrid

[16] Edel JB, Schulmann K, Skrzypek E, Cocherie A. Tectonic evolution of the European Variscan belt constrained by paleomagnetic, structural and anisotropy of magnetic susceptibility data from the northern Vosges magmatic arc (eastern France). Journal of the Geological Survey, London. 2015;**170**:785-804

[17] Faure M, Bé Mézème E, Cocherie A, Rossi P, Chemenda A, Boutelier D. Devonian geodynamic evolution of the Variscan Belt, insights from the French massif central and massif Armoricain. Tectonics. 2008;**27**:TC2005

[18] Tait JA, Bachtadse V, Franke W, Soffel HC. Geodynamic evolution of the European Variscan fold belt : Paleomagnetic and geological constraints. Geologische Rundschau. 1997;**86**:585-598

[19] Sousa MB. Litoestratigrafia e Estrutura Do «Complexo Xisto-Grauváquico» - Grupo Do Douro (Nordeste de Portugal). PhD thesis. University of Coimbra; 1982

[20] Dias R, Ribeiro A, Coke C, Pereira E, Rodrigues J, Castro P, Moreira N, Rebelo J. Evolução Estrutural Dos Setores Setentrionais Do Autoctone Da Zona Centro-Ibérica. 2013. Geologia de Portugal, vol. 1

[21] Pinto MD. Estudo comparativo de litologias calcosilicatadas e quartzo-pelíticas da Formação de Bateiras. Master thesis. University of Porto; 2016

[22] Silva A. A Litoestratigrafia e Estrutura do Supergrupo Dúrico – Beirão (Complexo Xisto-Grauváquico), em Portugal, e a sua correlação com as correspondentes sucessões em Espanha. Instituto Nacional de Engenharia, Tenologias e Inovação; 2005. pp. 2-26

[23] Silva PJABA. Mineralogia, Petrologia e geoquímica de Granitos e filões Aplito-pegmatíticos Da região de Guarda – Sabugal. PhD thesis. University of Trás-os-Montes e Alto Douro; 2014

[24] Sousa MB, Lowell GR. "Skarns" (W-F) Da auréola Do maciço granítico de Tabuaço (NE Portugal). III Congresso Nacional de Geologia. Sociedade Geológica de Portugal. Coimbra. 1991

[25] Cerejo T, Santos JF, Sousa JC, Castanho N, Sérgio G, Ribeiro S. Geochronology and isotope geochemistry of granitoids and metasediments of São Pedro das Águias area (Tabuaço W project). Comunicações Geológicas, 101. Especial I. 2014:61-64

[26] Ferreira N, Macedo CR, Sousa MB. Cronostratigrafia dos granitos da região de moimenta da beira – Tabuaço – Penedono. Memórias do Museu Laboratório Mineralógico e Geológico da Universidade do Porto, 1. 1987. pp. 287-301

[27] António T, Cerejo R. Geoquímica Da área de São Pedro Das Águias – Concessão de Tabuaço. University of Aveiro; 2013. Master thesis

[28] Dawson KM. Skarn Tungsten. Geology of Canadian Mineral Deposit Types, Geological Survey of Canada, Geology of Canada, No. 8. 1996. pp. 495-502

[29] Einaudi MT, Meinert LD, Newberry RJ. Skarn deposits: Economic Geology 75th Anniversary Volume. 1981:317-391

[30] Kwak TAP. W-Sn Skarn deposits and related metamorphic Skarns and Granitoids. Developments in Economic Geology. 1987;**24**:1-451

[31] Sato K. Characteristics of tungsten skarns in Japan, two contrasting types. Symposium on Tungsten Geology Jiangxi, China. 1982:203-210

[32] Sánchez SMT, Benito MCM, Pérez MLC. Mineralogical and physiochemical evolution of the Los Santos Scheelite skarn, Salamanca, NW Spain. Economic Geology. 2009;**104**:961-995

[33] Wesolowski D. Geochemistry of Tungsten in Scheelite Deposits: The Skarn Ores at King Island, Tasmania. Ph.D. Thesis. Pennsylvania State University; 1984

[34] Newberry RJ. W- and Sn-skarn deposits: A 1998 status report: Mineralogical Association of Canada Short Course Series. 1998;**26**:289-335

[35] Meinert LD, Dipple GM, Nicolescu S. World Skarn deposits. Economic Geology 100th Anniversary Volume. 2005:299-336

[36] Korzhinskii DS. Theory of Metasomatic Zoning (Trans. Jean Argrell). New York, Oxford: Clarendon Press; 1970. p. 162

[37] Newberry RJ, Einaudi MT. Tectonic and geochemical setting of Tectonic and Ore Deposits. Tucson, Ariz. March 19-20, Proc, 1981

[38] Meinert LD. Application of Skarn deposit zonation models to mineral exploration. Exploration and Mining Geology. 1998;**6**(2):185-208

[39] Mathieson GA, Clark AH. The Cantung E-zone scheelite skarn ore body, N.W.T.: A revised genetic model. Economic Geology. 1984;**79**(5):883-901

[40] Bird DK, Schiffman P, Elders WA, Williams AE, McDowell SD. Calc-silicate mineralization in active geothermal systems. Economic Geology. 1984;**79**:671-695

[41] Frisch CJ, Helgeson HC. Metasomatic phase relations in dolomites of the Adamello alps. American Journal of Science. 1984;**284**:121-185

Chapter 5

Lead Isotopes as Tracers of Metal Sources and Timing of the Carbonate-Hosted Pb-Zn Deposits in the Nappes Zone, Northern Tunisia

Nejib Jemmali and Fouad Souissi

Additional information is available at the end of the chapter

http://dx.doi.org/10.5772/intechopen.72690

Abstract

The polymetallic deposits in the Nappes zone, Northern Tunisia occur in the contact between Triassic-Miocene/Eocene carbonate rocks and in the Upper Cretaceous limestones. They can be divided into two groups: one is Pb-Zn mineralization with Hg and As in fractures with local intrusions of Neogene volcanics (e.g., Fej Hassene, Oued Maden), and the second is stratabound and karst Pb-Zn mineralization with arsenic and antimony hosted in the continental Neogene strata or situated immediately below them (Aïn Allega, Sidi Embarek, Jebel Hallouf-Sidi Bou Aouane, Bazina, Jalta and Jebel Ghozlane). Pb-isotopic compositions of galenas display a homogeneous Pb isotope signature. Generally, Pb isotope ratios on ores from the Jalta, Jebel Ghozlane, Jebel Hallouf, Oued Maden and Fedj Hassene plot between samples of the Late Miocene igneous rocks and the sedimentary country rocks of the Nefza area and between the upper crust and orogen curves. This intermediate position may imply potential mixing between end-member sources. Because the Pb-Zn mineralization is fault-controlled and spatially associated with the post-nappe Miocene series and the calculated model age is about 10.86 Ma, one is led to argue that the mineralization in the Nappes zone deposits occurred during the last paroxysmal phase of the Alpine folding (i.e., Miocene age).

Keywords: Pb isotopes, carbonate-hosted Pb-Zn deposits, mixing sources, Upper Miocene age, Nappes zone

1. Introduction

The Nappes zone (**Figure 1**), which constitutes the eastern prolongation of the Atlas orogenic belt of North Africa, is composed of thrust sheets that resulted from a major Neogene tectonic

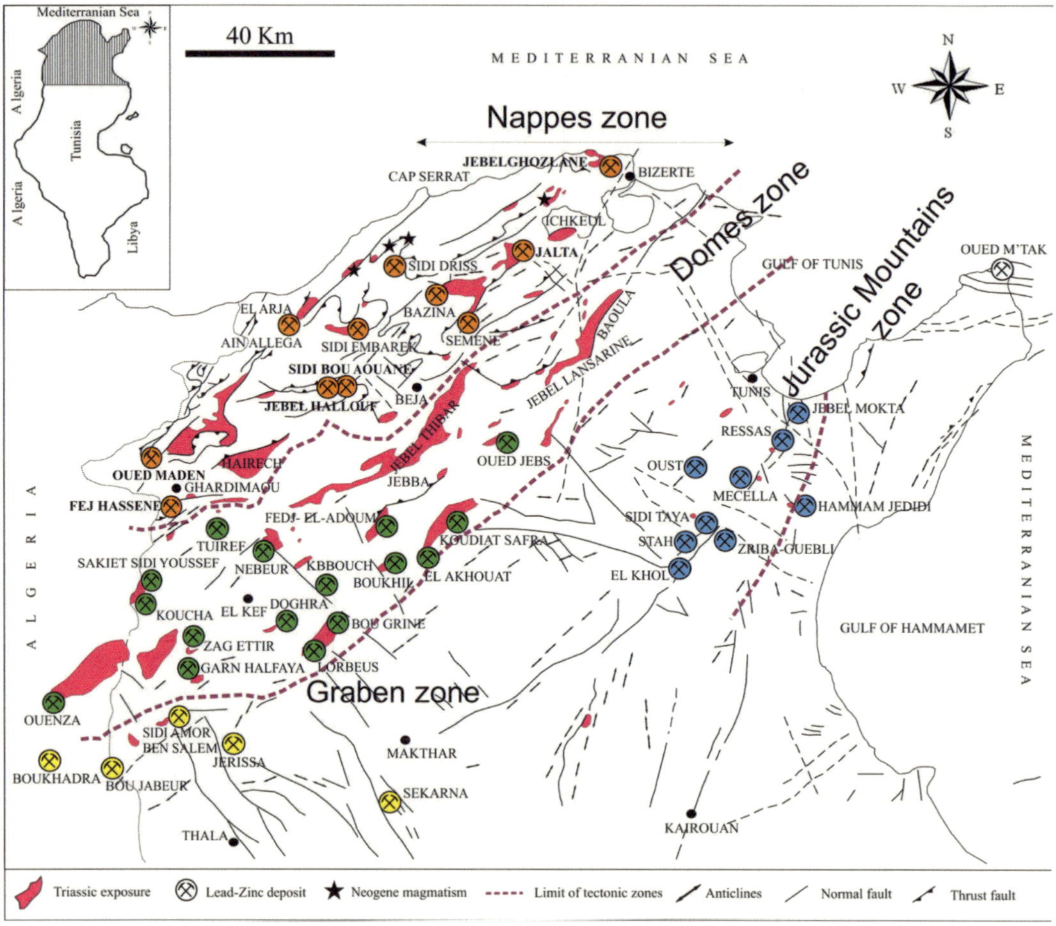

Figure 1. Geological zones, Triassic exposures and Pb-Zn deposits in Northern Tunisia (adapted from Jemmali et al. [1].) Brown circle: Nappes zone Pb-Zn-Hg (As) deposits. Green circle: Domes zone Pb-Zn (Ba-Sr) deposits. Yellow circle: Reef Aptian zone Pb-Zn deposits. Blue circle: F-Ba (Pb-Zn) deposits.

event [2]. The thrust sheets moved during the Early to Middle Miocene from NW to SE along regional westerly dipping fault planes in shear zones that vary in strike from E-W to SW-NE [2, 3]. The Nappes zone comprises Mesozoic and Tertiary sediments and is characterized by a major thrust sheet of Oligo-Miocene "Numidian" flysch that overrides younger flysch strata mainly of Cretaceous and Eocene age [4]. This zone includes the following units: (1) the Numidian nappe consists of a thick series of siliciclastic flysches made up of sandstones and argillites, which is of Oligocene to Burdigalian age; (2) the Tellian units dated Late Cretaceous to Early Eocene. It is composed of three formations: the Abiod (marl and limestones), El Haria (marl and shale) and Metlaoui (limestones); and (3) the Miocene-Pliocene Molasse includes coarse conglomerates and red sandy shales [4]. The Late Miocene major orogeny resulted in the Neogene basins along NNE-SSW directions and the emplacement of the nappes. This compressive event was followed by the development of longitudinal faults and bimodal volcanism [5, 6]: (1) granodiorite and rhyodacites were likely emplaced in a Serravallian-Tortonian compressive context, and (2) basalts may relate to the following Messinian rifting event. In the Nappes zone, the Triassic outcrops (variegated clays, sandstones, dolostones, gypsum breccias) often occur at the base of the overlapping units (Aïn Jantoura, Hédil), as well as diapiric units cross-cutting the allochthonous formations (structural trend of Ghardimaou-Ain Draham-Cap Serrat, Jebel Zouza) [7].

The Nappes zone hosts numerous polymetallic deposits (**Figure 1**). The mineralization is associated either with the Late Miocene orogeny (emplacement of the nappes) or with the Neogene volcanism. Two groups of deposits are present there: (1) stratabound and karst Pb-Zn mineralization with arsenic and antimony hosted in the continental Neogene strata or situated immediately below them (Aïn Allega, Sidi Embarek, Jebel Hallouf-Sidi Bou Aouane, Bazina, Semene, Jalta, Bechateur, Ain el Bay, Chouichia) and (2) Pb-Zn mineralization with arsenic and mercury in fractures with local intrusions of Neogene volcanics (Fej Hassene, Oued Maden, Jebel Arja).

The current study represents a synthesis of Pb isotope data collected from papers published by Jemmali et al. [1, 8–10] of Jalta, Jebel Ghozlane, Jebel Hallouf, Oued Maden and Fedj Hassene deposits in the Nappes zone (**Figure 1**) and comparison to previously published Pb isotope ratios of Neogene igneous rocks and sedimentary cover rocks from the same zone. This will help constrain the source(s) of metals and the possible age of mineralization.

2. Setting and characteristics of the Nappes zone Pb-Zn deposits: case study

2.1. Jalta

The lithostratigraphic units consist mainly of (**Figure 2**): (1) the Triassic series consist of chaotic shales and dolomitic limestones containing gypsum and alunite; (2) the Upper Cretaceous and Eocene series are composed of marls and limestone, respectively; (3) Middle to Upper Miocene strata are represented by continental, detrital facies composed of gray conglomerates and lenticular limestones, lacustrine limestones, and blue marls alternating with conglomerates, overlying the Triassic, Cretaceous and Eocene deposits; (4) the continental Pliocene series lies on the Upper Miocene conglomerates; and (5) the Quaternary continental deposits are represented by alluvium.

At the regional scale, the location of the basin of Jalta is mainly controlled by two main structural trends [12]: N30 Ras El Korane-Thibar major fault on the northwest side and Messeftine N-S fault on the eastern side accompanied on the southern side by the secondary N140E Mateur fault. At the deposit scale, the main structural features in and around the Jalta district consist mostly of strike-slip faults and few normal faults [4, 13], having different orientations (**Figure 2**). NE-SW trending sinistral strike-slip faults with N40-N60 exist far from the Jalta mine. In the northwestern part of the district, NE-SW trending sinistral strike-slip faults are considered as a deep fault where Triassic rocks are in contact with Miocene series. NW-SE trending dextral strike-slip faults with N120-N150, with distensive components, cut the Jalta

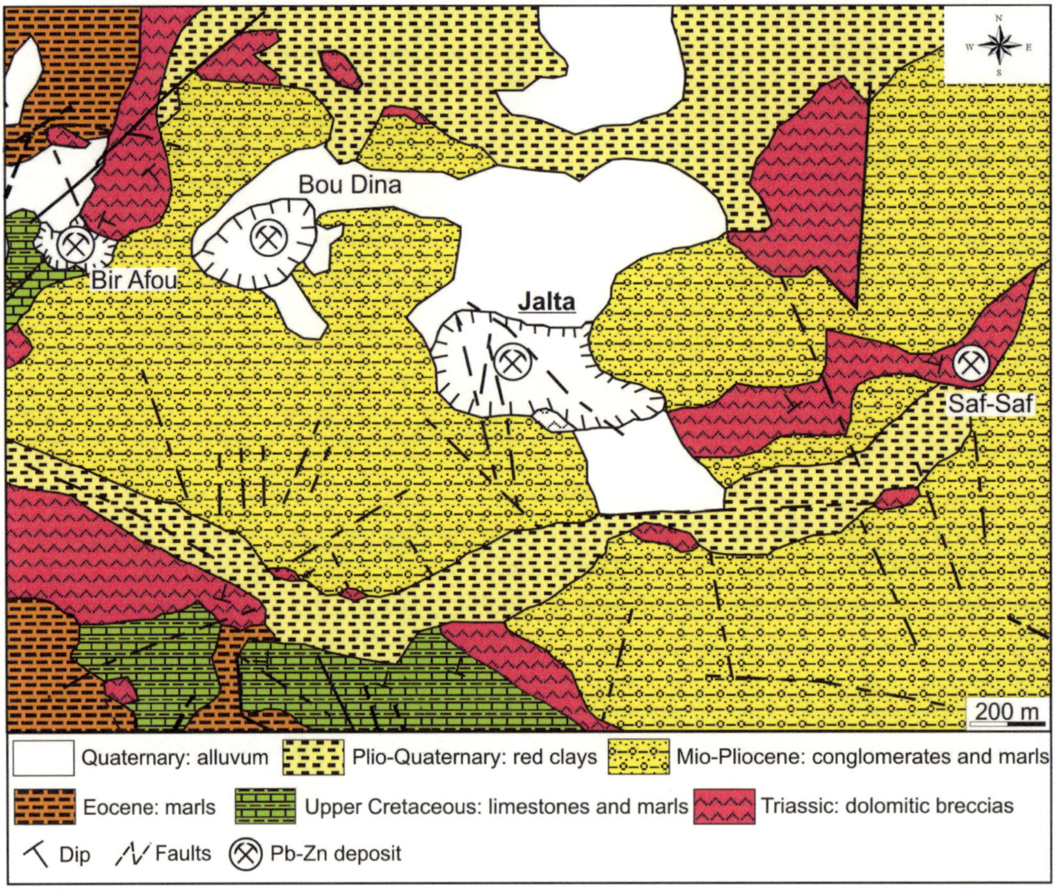

Figure 2. Geological map of Jalta district (adapted after Crampon [11].).

mine and other areas around it. These sinistral and dextral strike-slip faults host mineralization. In contrast, E-W trending normal faults, with a tectonic style of horst-and-graben structures, are more expressed outside west of the district and do not host Pb-Zn mineralization. All these sets of faults, especially the sinistral strike-slip faults, are due to the N-S directed deformation that affected the region during Miocene [4].

The Jalta mine has produced 75,000 tons of ore grading 59% Pb. The Pb-Zn mineralization in the Jalta deposit ([1, 10]; **Figure 2**) is located mainly near and along the contact of the Triassic rocks with the Miocene series. The mineralization mainly hosted by brecciated Triassic dolostones is composed of galena, barite and minor pyrite, jordanite and sphalerite occurring as veins, stockworks, disseminated, karstic and breccias cement [10].

2.2. Jebel Ghozlane

Based on several studies [14, 15], the stratigraphy of Jebel Ghozlane consists mainly of (**Figure 3**): (1) the Triassic breccias composed of variegated clays, sandstones and dolostones; (2) the Upper

http://dx.doi.org/10.5772/intechopen.72690

Cretaceous rocks consist of massive limestones with marl intercalations; (3) the Maastrichtian-Paleocene is represented by a marl-bearing series; (4) the Lower Eocene rocks consist of marine limestones with *Globigerina* sp.; (5) the Upper Eocene consists of marls and (6) the Quaternary is represented by marine facies. NNE-SSW trending dextral strike-slip fault characterizes the contact between the Triassic and Eocene at Jebel Ghozlane, but in other parts of the area the main contact between them is an unconformity. At Jebel Daouda, thrust faults trending NE-SW separate the Upper Eocene rocks from the Cretaceous and Triassic rocks, and the Upper Eocene rocks from the Lower Eocene rocks at Jebel Touila.

The Jebel Ghozlane deposit, which is among the largest of the Pb-Zn deposits in and near the Bechateur district, is situated along faults and a thrust-sheet boundary (**Figure 3**). It produces ca. 6680 tons of Pb and 53,128 tons of Zn pure metals. The deposit is hosted by Triassic dolostones and Lower Eocene dolomitic limestones. The orebodies, which occur as vein, dissemination and breccia styles, are localized along N150-160 dextral strike slip-fault contact between the Triassic and Eocene rocks (**Figure 2**). The ore of the deposit consists of galena, sphalerite and minor pyrite, with barite and celestite as gangue minerals.

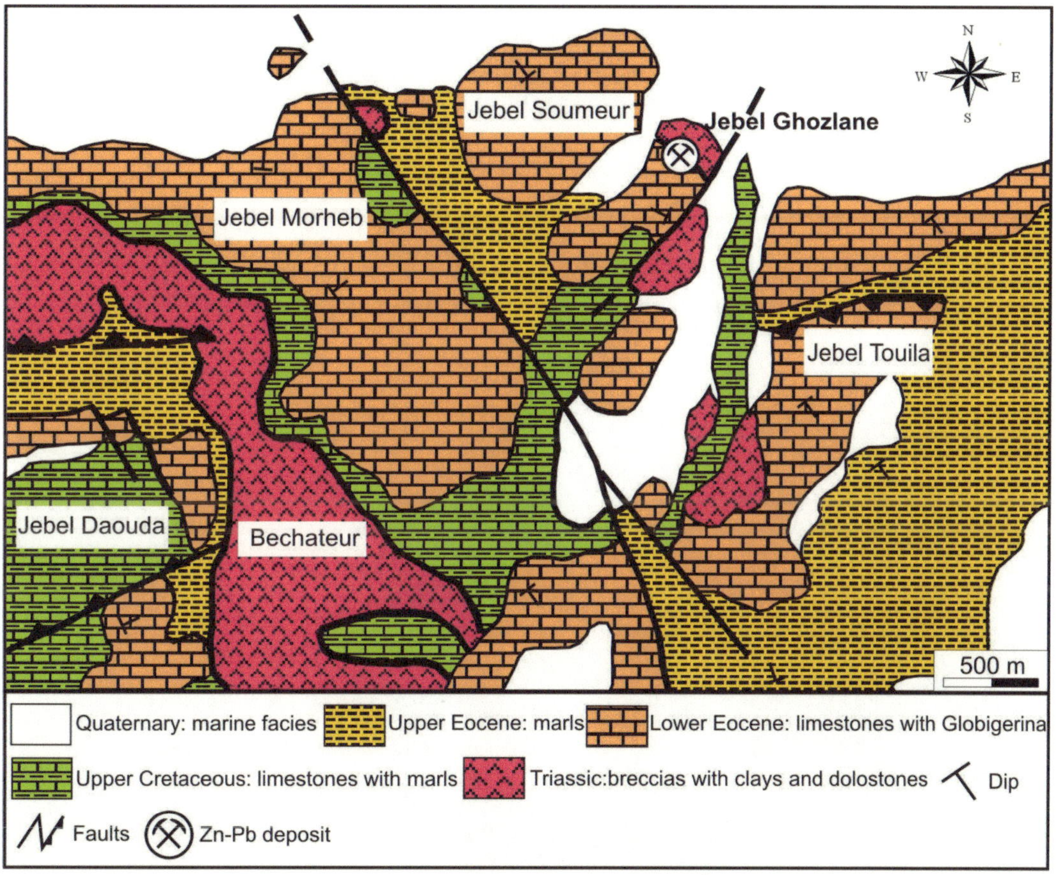

Figure 3. Geological map of Jebel Ghozlane. (Adapted from Melki et al. [15].)

2.3. Jebel Hallouf

The stratigraphy of Jebel Hallouf consists of a Triassic-Miocene series overlain in fault contact by the Kasseb Paleocene-Oligocene allochthonous unit, which in turn is overlain by the Neogene post-nappe continental series [4, 16] (**Figure 4**). The Triassic series consists of dolomitic breccia and evaporates, unconformably overlain by Lower-Upper Cretaceous clay-rich limestones. Upper Cretaceous-Paleocene series consists mainly of marls, which in turn are overlain by Eocene limestones. The overlying Oligocene-Miocene strata are made of sandstones. Five successive extensional and compressional episodes have characterized the tectonic evolution of Jebel Hallouf [17]: NE-SW-trending upright isoclinal open to closed folds, and a succession of closed-to-tight N30°–40°E oriented saline rock-cored anticlines and broad E-W to N-S-trending open-to-gentle synclines, truncated and/or accompanied by a series of ENE-WSW, NW-SE, E-W and N-S trending faults.

The Jebel Hallouf deposit was mined by the SOTEMI mining company between 1965 and 1986 and has produced since its exploitation in 1910 the equivalent of 326,541 tons of Pb and 14,207 tons of Zn metals. Mineralization is hosted by Campanian-Maastrichtian limestones. The style of mineralization consists of cavity-filling of karstic features that cut the mentioned host rocks on the north side of the Jebel Hallouf anticline. Vertical cavities developed from WSW-ENE and NNW-SSE joints were partially filled up by stratified mineralized calcite. The mineral association consists of calcite, sphalerite, galena, jordanite and pyrite.

2.4. Oued Maden

The lithostratigraphic units consist of a series of stacked nappe structures produced by regional westward intra-Miocene tangential over-thrusting and Upper Miocene to Pliocene-Quaternary post-nappe tectonic phases [18]. Two main nappes are recognized [4]: (1) the

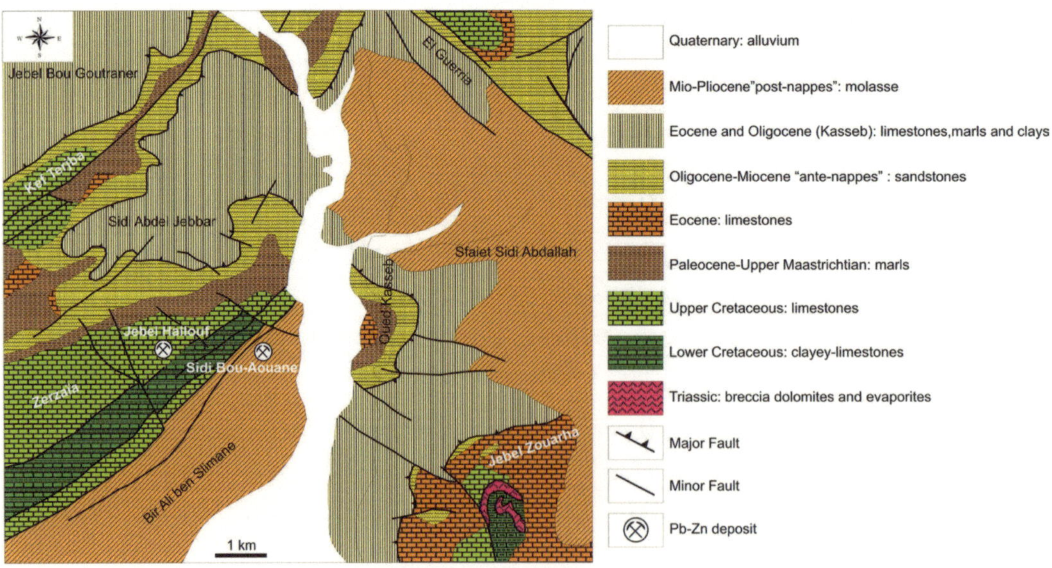

Figure 4. Geological map of Jebel Hallouf (adapted from [16].).

Numidian nappe consisting of Oligocene clay-rich sandstones and (2) the Ed-Diss nappe comprising Senonian (Upper Campanian-Lower Maastrichtian) and Eocene sediments. The autochthon beneath these nappes displays a normal stratigraphic sequence consisting of Santonian-Maastrichtian (Late Cretaceous) rocks (**Figure 5**). The Upper Santonian-Lower Campanian consists of thinly bedded gray marls with rare intercalations of marly limestones. These series are in fault contact with the underlying Triassic rocks and are overlain by thick succession of marls and limestones of Upper Campanian-Lower Maastrichtian age, followed by thick gray sulfide-rich limestones of Middle Maastrichtian age.

From 1900 to 1955, about 11,500 tons of Pb and 89 tons of Cu pure metals (with ca. 350 g/t Ag) were produced from open pit and underground workings [18]. Mineralization occurs as open-space fillings of veinlets and stockwork structures superimposed on the NE-SW-trending Groura and Ferza faults cutting across the Triassic dolostone and the Campanian-Maastrichtian limestone country rocks. The Groura and Ferza faults (**Figure 5**) are part of a regional major structure referred to as the Ghardimaou-Cap Serrat fault (**Figure 1**). The ore mineralogy is dominated by galena, sphalerite, pyrite and sulfosalts (mainly tetrahedrite).

2.5. Fedj Hassene

The lithostratigraphic units of Fedj Hassene (**Figure 6**) consist of: (1) Triassic series consist of limestone, dolostone and gypsum-bearing red argillite breccia, locally truncated by regional E-W striking faults; (2) Aptian gypsum-bearing dolomitic breccias; (3) Albian marls and limestones; (4) Cenomanian limestones and marls capped by a Turonian succession of marls and limestones; (5) Eocene nummulitic limestones; (6) Oligocene marls and sandstones; and (7) continental Neogene marls, sandstones and conglomerates. The Cretaceous ends with thick Coniacian-Santonian marls followed by a succession of Campanian-Maastrichtian marls and limestones. The major structure in the Fedj Hassene deposit is the Ain el Kohla ESE-WNW-trending fault.

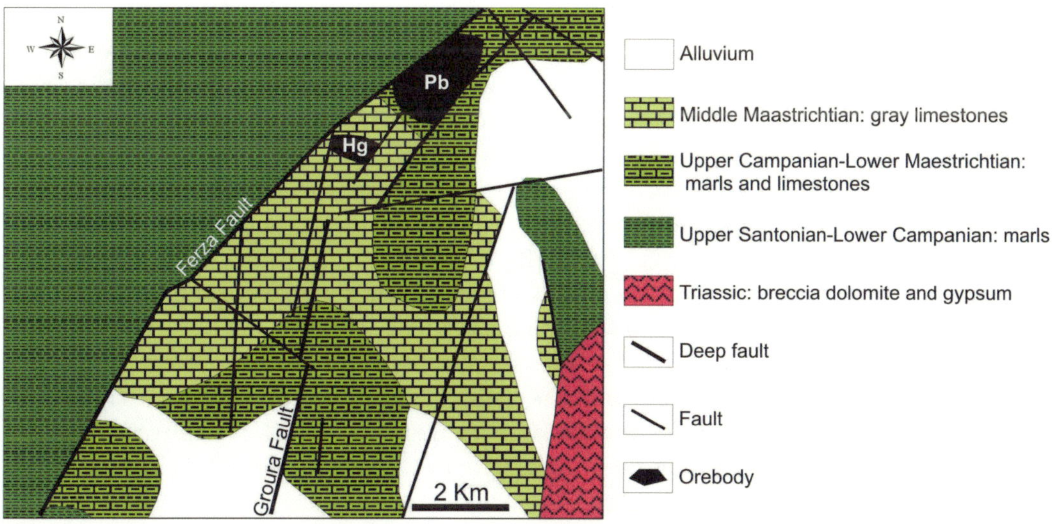

Figure 5. Geological map of Oued Maden. (Adapted from Slim-Shimi [19].)

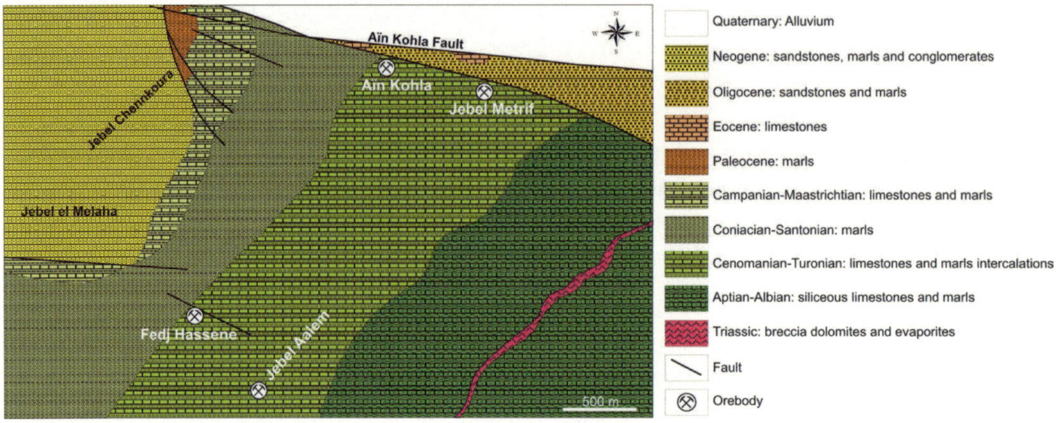

Figure 6. Geological map of Fedj Hassene. (Adapted from Sainfeld [20].)

During the lifetime of the mine that exploited the Fedj Hassene deposit until 1992, about 55,600 tons of Zn and about 300 tons of Pb pure metals have been produced. The Zn-Pb mineralization mainly occurs as ESE-WNW-trending veins and stockworks enclosed in Upper Cretaceous limestones. The mineralization consists mainly of sphalerite and galena with minor amounts of pyrite and chalcopyrite.

3. Data source

The lead isotopic compositions obtained on galenas from the abovementioned deposits are listed in **Table 1** and plotted on conventional covariation diagrams in **Figures 7** and **8**. The Pb isotope ratios range between 18.695 and 18.894 for $^{206}Pb/^{204}Pb$, 15.661 and 15.684 for $^{207}Pb/^{204}Pb$, and 38.718 and 38.917 for $^{208}Pb/^{204}Pb$. When it is plotted on conventional isotopic diagrams, Pb in the ore samples from the two deposits (Jalta and Jebel Ghozlane) defines two distinct fields (**Figures 7** and **8**). However, the radiogenic nature of the Pb characterizes the Oued Maden. The ore samples are quite homogeneous for each deposit and have higher $^{206}Pb/^{204}Pb$ values than samples from the Late Miocene igneous rocks of the Nefza area. The $^{207}Pb/^{204}Pb$ versus $^{206}Pb/^{204}Pb$ ratios plot between the orogene and upper crustal curves but close to the upper crustal reservoir in the plumbotectonics model of Zartman and Doe [22] (**Figure 7**), whereas the $^{208}Pb/^{204}Pb$ ratios versus $^{206}Pb/^{204}Pb$ plot slightly above the orogen curve. The narrow range of data may suggest a well-mixed source (see discussion). The compositional variation of Jalta and Jebel Ghozlane (two distinct fields), Jebel Hallouf, Oued Maden and Fedj Hassene may be due to an input of Pb from different sources and mixing of multiple metal-bearing brines. The samples from Jebel Hallouf, Jalta, and Jebel Ghozlane suggest ore deposition from a similar hydrothermal fluid. A comparison of the Pb isotope ratios of ore samples in the studied deposits has been made with previously published Pb isotope values of Late Miocene igneous rocks and sedimentary country rocks from the Nefza area (**Figure 7**). Some of the ore samples have Pb isotope ratios that plot very close to previous results by Decrée et al. [21].

Deposit name	$^{206}Pb/^{204}Pb$	$^{207}Pb/^{204}Pb$	$^{208}Pb/^{204}Pb$
Jalta	18.8210	15.6770	38.8740
	18.8230	15.6760	38.8750
	18.7200	15.6650	38.7300
	18.8200	15.6740	38.8708
	18.8510	15.7290	39.0610
Oued Maden	18.8900	15.6800	38.9070
	18.8940	15.6840	38.9170
	18.8840	15.6750	38.8880
Fedj Hassene	18.7920	15.6660	38.8490
	18.7880	15.6630	38.8620
	18.7880	15.6630	38.8380
Jebel Hallouf	18.8200	15.6780	38.9140
	18.8200	15.6800	38.9130
	18.8200	15.6760	38.8690
Jebel Ghozlane	18.7140	15.6680	38.7250
	18.7030	15.6680	38.7430
	18.7020	15.6670	38.7390
	18.7020	15.6670	38.7320

Table 1. Lead isotope composition of galena from Jebel Ghozlane [1, 8], Jalta [1, 10], Oued Maden, Fedj Hassene and Jebel Hallouf [8].

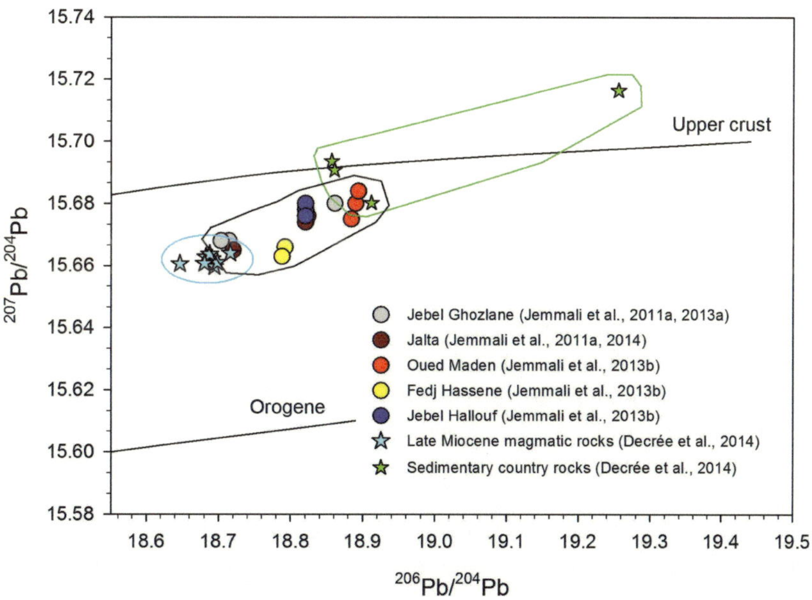

Figure 7. Plots of $^{207}Pb/^{204}Pb$ vs. $^{206}Pb/^{204}Pb$ for the Pb-Zn deposits in Jalta, Jebel Ghozlane, Jebel Hallouf, Oued Maden and Fedj Hassene compared to the Late Miocene igneous rocks and sedimentary country rocks from the Nefza area [21]. Curves of growth trends for Pb isotope ratios are from the plumbotectonic model of Zartman and Doe [22].

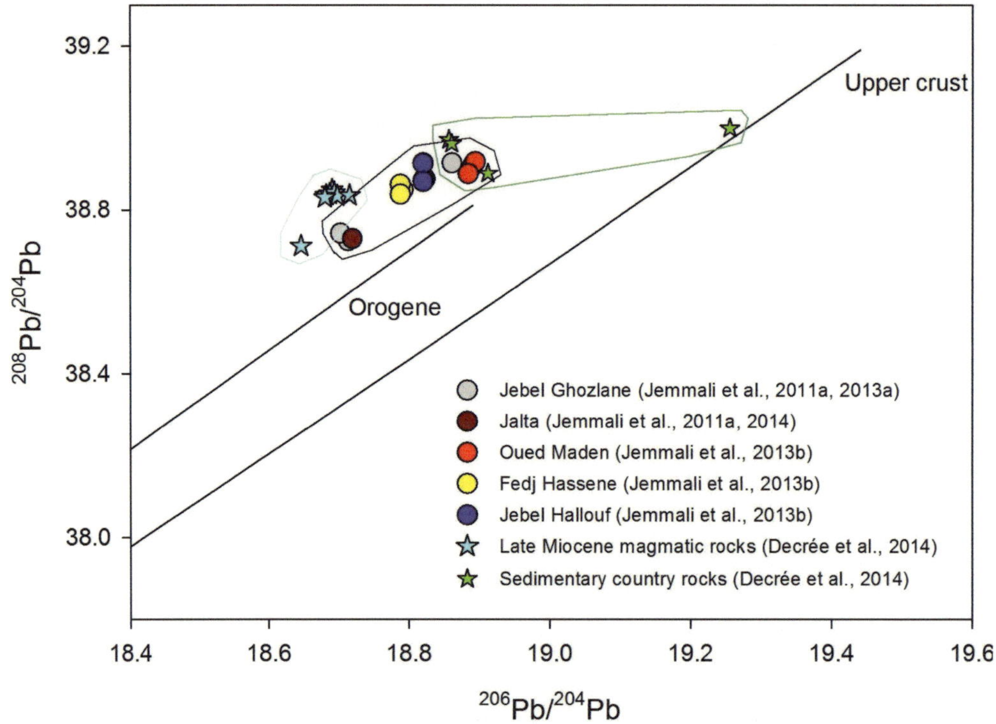

Figure 8. Plots of $^{208}Pb/^{204}Pb$ vs. $^{206}Pb/^{204}Pb$ for the Pb-Zn deposits in Jalta, Jebel Ghozlane, Jebel Hallouf, Oued Maden and Fedj Hassene compared to the Late Miocene igneous rocks and sedimentary country rocks from the Nefza area [21]. Curves of growth trends for Pb isotope ratios are from the plumbotectonic model of Zartman and Doe [22].

4. Discussion

4.1. Source(s) of metals

The restricted range of isotopic ratios in the studied ore deposit cluster may suggest that Pb originated from a well-mixed source. Overall, the Pb isotope ratios of ore samples plot within the field defined by Late Miocene igneous rocks and sedimentary country rocks implying potential mixing between end-member sources originating from the abovementioned similar metal source(s) (**Figure 7**). The Late Miocene igneous rocks correspond to the basement of the Atlasic foreland from which they are originated [23]. The plausibility of basement rocks as the deep-seated source of Pb in the studied ore deposits is supported by the similarity of Pb isotopic ratios between the Late Miocene igneous rocks and the abovementioned deposits, as well as the presence of jordanite, orpiment and realgar minerals. Jordanite is also present in the Lengenbach Pb-Zn-As-Tl-Ba deposits, which are hosted in Triassic dolostones in the Swiss Alps, for which it was proven that Pb and other metals were leached from basement rocks [24]. Another support for the plausibility of basement rocks as the deep-seated source of radiogenic Pb in the studied deposits is the presence of inherited faults connected to deep-seated faults cutting the Mesozoic-Cenozoic cover and the Triassic salts (**Figure 3**) and possibly extending into the basement, like the NE-SW Ghardimaou-Cap Serrat, the NE-SW Ras El Korane-Thibar and the N-S basement faults. However, the occurrence of the studied deposits mainly in Upper Cretaceous limestones and Miocene conglomerates also suggests plausible late remobilization of metals.

Nevertheless, the positions of the $^{208}Pb/^{204}Pb$–$^{206}Pb/^{204}Pb$ data slightly above the orogen curve (**Figure 8**) imply some contribution of Th-derived Pb (cf. [25]). The likely source of Th-derived Pb is the rocks in the Hercynian fold belt of North Africa, which include monazite-bearing Paleozoic metasediments and Hercynian granitoids [26]. These Paleozoic rocks are host to some of the world's economically significant base metal deposits [27], have been deformed and/or emplaced during the Hercynian orogeny, and form the basement exposed in scattered inliers of the Atlas systems in Morocco and Algeria [28, 29]. Thus, it is plausible that basement rocks in Northern Tunisia are similar to those in Algeria and Morocco. If that is the case, and following the plumbotectonic models [22], the source of Pb in galena of the studied deposits can be inferred as follows. Lead in the northern African Paleozoic basement rocks, which were deformed and/or emplaced during the Hercynian orogeny, became part of a well-mixed multi-source upper crust reservoir. Evidence to this is the rather homogenous Pb isotope data from Zn-Pb ores hosted in pre-Hercynian metasediments and from Hercynian granitoids [27]. Then, Pb in the well-mixed multi-source upper crust reservoir was partly recycled back into the mantle and/or passed on to favorable host rocks in an orogeny [22, 30]. The orogeny in this case is the Maghrebides fold-thrust belt, which was formed during the Alpine orogeny and where the studied Pb-Zn deposits are hosted primarily in Triassic carbonate rocks and partly also in the cover rocks. A second source may be the sedimentary cover rocks, and the Pb isotope ratios of the studied ore deposits are plotting near to the Pb isotope values defined by the second source (**Figure 7**). The current Pb isotope data support Decrée et al. [23] conclusion, suggesting that mixing between end-member sources originating from the basement igneous rocks and the sedimentary cover rocks was responsible for providing metals for the Nappes zone deposits.

4.2. Age of mineralization

The governing mechanisms during the first phase of halokinesis which has taken place likely during Jurassic to Middle Cretaceous [31] were presumably controlled by Jurassic-Lower Cretaceous normal faults inherited from Tethyan rifting [32]. The salt diapirs have been active again during the Alpine orogeny (Early-Middle Miocene compressional events produced folds, nappe emplacement and bimodal volcanism and restarted the halokinetic phenomena [31]). Because the Pb-Zn mineralization in the Nappes zone is associated with the post-nappe Miocene series, one is led to conclude, therefore, that the mineralization is related to the last paroxysmal phase of the Alpine folding (i.e., Miocene age). Because the Pb-Zn mineralization in the studied deposits is fault-controlled and spatially associated with the post-nappe Miocene series, one is led to hypothesize that the mineralization occurred during the last paroxysmal phase of the Alpine folding (i.e., Miocene age). The calculated model ages using the Pb isotope model of Stacey and Kramers [33] range from 2.7 to 21.6 Ma with a median of 10.86 Ma (excluding negative values of Oued Maden), indicating an Upper Tertiary-Quaternary age. This age, which was attributed to the F-(Ba-Pb-Zn) ores of the Zaghouan district [14, 34, 35] and to the polymetallic mineralization of the Nappes zone [23, 36] and which is similar to that recently proposed for the world-class Touissit-Bou Beker district of northeastern Morocco [37], coincides with the Serravallian-Tortonian magmatism event [38] and with the mid-Miocene Alpine compressional tectonics in Northern Tunisia [39]. Thus, the Tertiary orogeny in this case, as mentioned before, is represented by the Maghrebides fold-thrust belt, which was formed during the Alpine orogeny. Similarly, a genetic link between the Messinian mafic magmatism and Sidi Driss Pb-Zn deposit has been proposed by Decrée et al. [40]. In the Oued Maden deposit, the Pb-isotopic data give negative model ages. This implies that an anomalous Pb was introduced to characterize ore Pb in the Oued Maden deposit, which gave negative or excess model age suggesting significant radiogenic contamination. This kind of highly radiogenic Pb known as J-type Pb [41] could be derived from high U and Th crustal source reservoirs. Accordingly, resulting radiogenic Pb was likely stored in basement rocks before being remobilized and redeposited in the cover rocks in the studied area.

5. Conclusions

This study presents a lead isotope database of 18 galena samples from the selected deposits belongs to the Nappes zone, Northern Tunisia. The isotopic signatures of these carbonate-hosted Pb-Zn deposits reflect the source(s) of metals and the probable age of mineralization. The current Pb isotope data support suggests that mixing between end-member sources originating from the basement igneous rocks and the sedimentary cover rocks was responsible for providing metals for the Nappes zone deposits. The calculated model indicates an Upper Tertiary-Quaternary age for the emplacement of the mineralization. This study may be useful for mineral exploration and archaeological correlation of metal artifacts.

Author details

Nejib Jemmali[1*] and Fouad Souissi[2]

*Address all correspondence to: nejib.jemmali@yahoo.fr

1 Faculté des Sciences de Gafsa, Universite de Gafsa, Gafsa, Tunisia

2 Département de Géologie, Faculté des Sciences, Université de Tunis El Manar, Tunis, Tunisia

References

[1] Jemmali N, Souissi F, Villa IM, Vennemann T. Ore genesis of Pb–Zn deposits in the Nappes zone of Northern Tunisia: Constraints from Pb–S–C–O isotopic systems. Ore Geology Reviews. 2011a;**40**:41-53

[2] Ben Ayed N. Evolution tectonique de l'avant pays de la chaîne alpine de Tunisie [thesis]. France: Université de Paris Sud France; 1986. 328 p

[3] Wildi W. La chaîne tello-rifaine (Algérie, Maroc, Tunisie): Structure, stratigraphie et évolution du Trias au Miocène. Revue de Géologie dynamique de géographie physique. 1983;**24**:201-297

[4] Rouvier H. Géologie de l'Extrême-Nord tunisien: tectoniques et paléogéographies superposées à l'extrêmité orientale de la chaîne nord-maghrébine [thesis]. Paris, France: Univ. Pierre et Marie Curie; 1977. 703 p

[5] Badgasarian GP, Bajanik S, Vass D. Age radiométrique du volcanisme néogène du Nord de la Tunisie. Notes de service géologique Tunisie. 1972;**40**:79-85

[6] Halloul N. Géologie, pétrologie et géochimie du bimagmatisme néogène de la Tunisie septentrionale (Nefza et Mogods). Implications pétrogénétiques et interprétations géodynamiques [thesis]. France: University of Clermont-Ferrand; 1989. 270 p

[7] Perthuisot V, Hatina N, Rouvier H, Steinberg M. Concentration métallique (Pb–Zn) sous un sur-plomb diapirique: Exemple de Jebel Bou Khil (Tunisie Septentrionale). Bull Société Géologique de France. 1987;**8**:1153-1160

[8] Jemmali N, Souissi F, Carranza EJM, Vennemann T. Mineralogical and geochemical constraints on the genesis of the carbonate-hosted Jebel Ghozlane Pb–Zn deposit (Nappes zone, Northern Tunisia). Resource Geology. 2013a;**63**(1):27-41

[9] Jemmali N, Souissi F, Carranza EJM, Bouabdellah M. Lead and sulphur isotope constraints on the genesis of the polymetallic mineralization at Oued Maden, Jebel Hallouf and Fedj Hassene carbonate-hosted Pb–Zn (As–Cu–Hg–Sb) deposits, Northern Tunisia. Journal of Geochemical Exploration. 2013b;**132**:6-14

[10] Jemmali N, Souissi F, Carranza EJM, Vennemann T, Bogdanov K. Geochemical constraints on the genesis of the Pb–Zn deposit of Jalta (northern Tunisia): Implications for timing of mineralization, sources of metals and relationship to the Neogene volcanism. Chemie der Erde. 2014;**74**:601-613

[11] Crampon N. Etude géologique générale de Djalta. A.G.M.I. Tunis: ONM; 1965. 48 p

[12] Melki F, Zouaghi T, Harrab S, Casas Sainz A, Bédir M, Zargouni F. Structuring and evolution of Neogene transcurrent basins in the Tellian foreland domain, north-eastern Tunisia. Journal of Geodynamics. 2011;**52**:57-69

[13] Roussev G, Raivoev B, Papov A. Gisement de plomb de Jalta. Rapport géologique avec estimation des réserves. Compagnie de recherche 1974-1975. Société Tunisienne d'expansion minière. Convention de renouvellement des réserves des mines en activité du 11-06-1974. Tunis: Technoexportstroy-Bulgareprorémi-Bulgarie; 1976. 101 p

[14] Jemmali N, Souissi F, Vennemann T, Carranza EJM. Genesis of the Jurassic carbonate-hosted Pb–Zn deposits of Jebel Ressas (North-Eastern Tunisia): Evidence from mineralogy, petrography and trace metal contents and isotope (O, C, S, Pb) geochemistry. Resource Geology. 2011b;**61**(4):367-383

[15] Melki F, Alouani R, Boutib L, Zargouni F, Tlig S. Notice explicative de la carte géologique de la Tunisie à 1/50.000, Bizerte, Feuille n°2; 2001

[16] Mansouri A. Gisement de Pb–Zn et karstification en milieu continental: le district minier de djebel Hallouf-Sidi bou Aouane (Tunisie septentrionale) [thesis]. Univ. P. et M. Curie, Lab. Géologie Appl; 1980. 257 p

[17] Charef A. La nature et le rôle des phases associées à laminéralisation Pb–Zn dans les formations carbonatées et leurs conséquences métallogéniques. Etude des inclusions fluides et des isotopes (H, C, O, S, Pb) des gisements des Malines (France), Fedj-el-Adoum et Jbel-Hallouf–Sidi Bou Aouane (Tunisie) [thesis]. France: University of Nancy; 1986. 291 p

[18] Gharbi M. Etude des minéralisations mercurifères de l'accident Cap Serrat–Ghardimaou (Tunisie du Nord–Ouest) [thesis]. l'institut national polytechnique de Loraine; 1977. 131 p

[19] Slim-Shimi N. Minéralogie et paragénése des gîtes plymetalliques de la zone des nappes en Tunisie. Conditions géochimiques de dépôt et implications génétiques [thesis]. Fac. Sci De Tunis; 1992. 268 p

[20] Sainfeld P. Les gîtes plombo-zincifères de Tunisie. Annales des Mines et de la Géologie. Tunis; 1952. 285 p

[21] Decrée S, Marigna C, Liégeois JP, Yans J. Miocene magmatic evolution in the Nefza district (Northern Tunisia) and its relationship with the genesis of polymetallic mineralizations. Lithos. 2014:240-258

[22] Zartman RE, Doe BR. Plumbotectonics: The model. Tectonophysics. 1981;**75**:135-162

[23] Decrée S, Marignac C, Abidi R, Jemmali N, Deloule E, Souissi F. Tectonomagmatic context of Sedex Pb–Zn and polymetallic ore deposits of the Nappes zone Northern Tunisia, and comparisons with MVT deposits in the region. In: Bouabdellah M, Slack JF, editors. Mineral Deposits of North Africa, Mineral Resource Reviews. Switzerland: Springer International Publishing; 2016. pp. 497-525

[24] Hofmann BA, Knill MD. Geochemistry and genesis of the Lengen-bach Pb–Zn–As–Tl–Ba-mineralisation, Binn Valley, Switzerland. Mineralium Deposita. 1996;**31**:319-339

[25] Tornos F, Chiaradia M. Plumbotectonic evolution of the Ossa Morena Zone, Iberian Peninsula: Tracing the influence of mantle-crust interaction in ore-forming processes. Economic Geology. 2004;**99**:965-985

[26] Tahiri A, Fernando Simancas J, Azor A, Galindo-Zaldívar J, González Lodeiro F, El Hadi H, Martínez Poyatos D, Ruiz-Constán A. Emplacement of ellipsoid-shaped (diapiric?) granite: Structural and gravimetric analysis of the Oulmès granite (Variscan Meseta, Morocco). Journal of African Earth Sciences. 2007;**48**:301-313

[27] Bouabdellah M, Beaudoin G, Leach DL, Grandia F, Cardellach E. Genesis of the Assif El Mal Zn–Pb (Cu, Ag) vein deposit. An extension-related Mesozoic vein system in the High Atlas of Morocco. Structural, mineralogical, and geochemical evidence. Mineralium Deposita. 2009;**44**:689-704

[28] Hoepffner C, Soulaimani A, Piqué A. The moroccan hercynides. Journal of Africa Earth Sciences. 2005;**43**:144-165

[29] Piqué A, Tricart P, Guiraud R, Laville E, Bouaziz S, Amrhar M, Ouali R. The Mesozoic–Cenozoic Atlas belt (North Africa): An overview. Geodinamica Acta. 2002;**12**:185-208

[30] Kramers JD, Tolstikhin IN. Two terrestrial lead isotope paradoxes, forward transport modelling, core formation and the history of the continental crust. Chemical Geology. 1997;**139**:75-110

[31] Rouvier H, Perthuisot V, Mansouri A. Pb–Zn deposits and salt-bearing diapirs in Southern Europe and North Africa. Economic Geology. 1985:666-687

[32] Khomsi S, Bédir M, Ben Jemia MG. Mise en évidence et analyse d'une structure atlasique ennoyée au front de la chaîne alpine tunisienne. Comptes Rendus Geoscience. 2004;**336**:1293-1300

[33] Stacey JS, Kramers JD. Approximation of terrestrial lead isotope evolution by a two-stage model. Earth and Planetary Science Letters. 1975;**26**:207-221

[34] Jemmali N, Carranza EJM, Zemmel B. Isotope geochemistry of Mississippi Valley Type stratabound F-Ba-(Pb-Zn) ores of Hammam Zriba (Province of Zaghouan, NE Tunisia). Chemie der Erde. 2017;**77**:477-486

[35] Souissi F, Jemmali N, Souissi R, Dandurand JL. REE and isotope (Sr, S, andPb) geochemistry to constrain the genesis and timing of the F–(Ba–Pb–Zn) ores of the Zaghouan district (NE Tunisia). Ore Geology Reviews. 2013;**55**:1-12

[36] Jemmali N. Le Trias du nord de la Tunisie et les minéralisations associées: minéralogie, géochimie (traces, isotopes O, C, S, Pb) et modèles génétiques [thesis]. Faculté des Sciences de Tunis; 2011. 255 p

[37] Bouabdellah M, Sangster DF, Leach DL, Rown AC, Johnson CA, Emsbo P. Genesis of the touissit-bou beker mississippi valley-type district (morocco–algeria) and its relationship to the Africa Europe collision. Economic Geology. 2012;**107**:117-146

[38] Jallouli C, Mickus K, Turki MM, Rihane C. Gravity and aeromagnetic constraints on the extent of Cenozoic volcanic rocks within the Nefza–Tabarka region, northwestern Tunisia. Journal of Volcanology and Geothermal Research. 2003;**122**:51-68

[39] Bouaziz S, Barrier E, Soussi M, Turki MM, Zouari H. Tectonic evolution of the northern African margin in Tunisia from paleostress data and sedimentary record. Tectonophysics. 2002;**357**(1):227-253

[40] Decrée S, Marignac C, De Putter T, Deloule E, Liégeois JP, Demaiffe D. Pb-Zn mineralization in a Miocene regional extensional context: The case of the Sidi Driss and the Douahria ore deposits (Nefza mining district, northern Tunisia). Ore Geology Reviews. 2008;**34**:285-303

[41] Evans AM. An Introduction to Economic Geology and Its Environmental Impact. Wiley-Blackwell; 1997. 376 p

Chapter 6

Trace Elements in Coal Gangue: A Review

Shaoqing Guo

Additional information is available at the end of the chapter

http://dx.doi.org/10.5772/intechopen.71335

Abstract

Coal gangue is one of the largest industrial residues. It has high ash content, low carbonaceous content, and heating value. Meanwhile, it has some trace elements. Large quantities of coal gangue cause serious environmental problems by polluting the air, water, and soil as well as occupying a tremendous amount of land. Now, coal gangue utilization is a matter of great concern and has attracted wide interest. However, some toxic trace elements in coal gangue should be paid more attention during the utilization of coal gangue. In this article, the modes of occurrence and the leaching characters of trace elements in coal gangue were introduced according to the result of the sequential extraction method and the leaching method. The release character of trace elements during combustion of coal gangue and the environmental implication of trace elements in coal gangue were also discussed. The sulfide-bound trace elements are dominant form in coal gangue. Leaching behavior of trace elements from coal gangue is affected by many factors. Different trace elements presented different transformation behaviors. Trace elements in coal gangue could release out and produce environmental implication in various degrees, depending on the type of trace elements.

Keywords: trace elements, coal gangue, modes of occurrence, release, environmental implication

1. Introduction

Coal gangue is a significant residue during the processing of coal mining and coal washing, which accounts for approximately 10–15% of raw coal production and becomes one of the largest industrial residues in China [1]. The rapid growth of the mining industry in China over the last decades has resulted in large amounts of coal gangue piled in wastelands. According to Zhao et al., there are 1500 coal waste gobs in China and 108 are located in Shanxi province [4]. It is estimated that about 40% of coal gangue gobs undergo spontaneous combustion in Shanxi province [3].

Coal gangue contains inorganic minerals and organic matter with high ash content, high pyrite content, and low heating value [1]. Besides the conventional pollutants of sulfur and nitrogen, it is enriched in some toxic elements which are present in less than 0.1% in coal gangue matrix [2–5]. Because of the existence of harmful trace elements in coal gangue, large quantities of coal gangue piled on the land can cause serious environmental problems by polluting the air, water, and soil as well as resulting in ecological risk [6–11]. Therefore, some reasonable methods should be taken to reduce the amount of coal gangue.

Recently, coal gangue utilization is a matter of great concern and has attracted wide interest in China [12]. Due to the lack of energy resources in China, coal gangue has been extensively utilized as a raw material for power plants to effectively utilize its calorific value [13, 14]. It is considered as a reasonable way to reduce the amount of coal gangue and environmental problems as well as bring some economic benefits [15]. According to the China resources utilization annual report 2014, the utilization of coal gangue as feed fuel for power plant was up to 150 million tons in 2013 in China [15]. However, during the utilization of coal gangue for power generation, the emissions of trace elements may have some severe environmental impacts besides the emission of sulfur dioxide, NOx or particulate matter [16]. It is well known that the trace element emissions during power generation may cause serious effects on human beings and the environment [15]. It is also reported that high emissions of trace element can cause some adverse effects for human health including acute and chronic lung injuries, inflammation and changed immunological mechanisms as well as increased cancer risk [17].

Over the past decades, extensive studies have been conducted on trace element in coal, especially trace element emissions from coal-firing power plants [18–21]. It is reported that the emission of trace elements is influenced by the thermodynamic properties of elements, the forms of elements in the feed fuel as well as the combustion environment [22–27]. Besides the experimental investigation, thermodynamic equilibrium calculation is also widely used to study the trace element behavior during coal combustion [28–32]. In addition, some studies have investigated the interactions between trace elements with the ash components and reported that the interactions play an important role in the emission behavior of trace element [33, 34]. Meanwhile, some studies such as the modes of occurrence and leaching behavior of trace element in coal have been conducted. Overall, the studies about the trace element in coal have provided meaningful information for the development of trace element control technologies.

In comparison to coal, coal gangue has a higher content of trace element and mineral [15]. Therefore, it can be inferred that the emission behavior of trace elements in coal gangue may be significantly different from that of coal, and thus the emission control of trace element from coal gangue is extremely important, especially from the combustion of coal gangue [15]. Although the combustion of coal gangue for power generation is rapidly developing, the studies on the trace element emissions are limited thus far. Some studies about the trace elements in coal gangue have been conducted, such as trace elements abundance properties, partitioning character or the release behavior during coal gangue spontaneous combustion [3, 4, 14, 35]. For the sustainable development of coal gangue utilization, the present work intends to provide information regarding the modes of occurrence of trace elements in coal gangue and the leaching character of trace elements from coal gangue. Also, the release character of trace element during combustion of coal gangue as well as the environmental implication of trace element in coal gangue is discussed in this chapter.

2. Modes of occurrence of trace elements in coal gangue

2.1. The mineral and chemical composition of coal gangue

Generally, the main mineral phases of coal gangue are kaolinite, quartz, pyrite, illite, calcite, ankerite, and montmorillonite [36, 37]. The coal gangue samples are usually dominated by silicon dioxide and aluminum oxide, followed by ferric oxide and potassium oxide accompanied with some trace elements [37]. For example, it is reported that the mineral phases in the Huainan coal gangue samples are kaolinite, quartz, illite, calcite, pyrite, siderite, and muscovite. The major chemical compositions in the samples are silicon dioxide, aluminum oxide, ferric oxide, and calcium oxide. Traces components such as potassium oxide, sodium oxide, magnesium oxide, and titanium dioxide are also identified [38]. The properties of the coal gangue samples from the Huainan Coalfield are presented in **Table 1** [38]. According to the mineralogical analysis, silicon, aluminum, potassium, and magnesium are mainly associated with the clay and quartz minerals. Calcium is mainly associated with carbonate minerals while iron is mainly associated with sulfide minerals, carbonate minerals, and the clay minerals [38].

Trace elements in coal gangue could be associated with the organic matter and inorganic matter in coal gangue matrix, but prefer to be associated with inorganic matrix, such as clay, sulfides, and carbonate minerals in coal gangue [39, 40]. In comparison with the background values, the content of the total trace element in coal gangue could provide important information about environment pollution [38]. Therefore, some researches regarding of the content of the total trace element in coal gangue have been conducted. For example, 12 trace elements,

Proximate analysis, dry basis (wt.%)							
Moisture		**Ash**		**Fixed carbon**		**Volatile matter**	
2.63		70.79		18.24		23.37	
Ultimate analysis (wt.%)							
C	**H**		**N**		**O**		**S**
20.14	2.07		0.27		9.87		1.21
Ash analysis (wt.%)							
SiO_2	**Al_2O_3**	**Fe_2O_3**	**CaO**	**MgO**	**K_2O**	**Na_2O**	**TiO_2**
54.18	20.40	9.28	13.60	0.18	0.44	0.64	1.33

Table 1. The chemical compositions of the coal gangue.

including As, V, Cr, Co, Ni, Cu, Zn, Se, Cd, Sn, Pb, and Hg, were selected by Yang et al. for the study of trace element in coal gangue at an open-cast coal mine area, Inner Mongolia, China [37]. Also, 20 trace elements, including Ba, Co, Cr, Cu, Ga, Mn, Ni, Li, Se, Zn, V, As, etc., were selected by Zhou et al. for the study of trace element in coal gangue at the Huainan Coalfield in China [38]. The comparison of trace element content in the coal gangue with black shales and Clarke values were used to evaluate the trace element enrichment by Zhou et al. [38]. It is reported that compared with the background values of black shales, the trace elements in coal gangue are enriched in Ga and depleted in Ba, Co, Cr, Cu, Li, Mn, Ni, Zn, V, Ce, Dy, Er, Gd, La, Lu, Nd, Yb, and Y, and the other trace elements are nearly equal to the background values of black shales [38, 41]. Meanwhile, the content of Ga, Se, As, and Sn are much higher in the Huainan Coalfield coal gangue compared with the Clarke values, while the content of other trace elements are lower [38]. The coal gangue in the Wulantuga open-cast coal mine area is enriched in Co and depleted in Se, Cd, and Hg, and the other trace elements are close to the concentrations of the world Clarke values [37].

2.2. The sequential extraction method

The sequential extraction method was organically employed to determine the mode of occurrence of toxic trace elements in soil samples and was later modified to evaluate the trace elements in coal [42]. This method identifies trace elements species on the basis of the chemical leaching of the complex substrate, which have been proved to be an effective method in determining the modes of occurrence of trace element in the solids [43, 44]. Based on sequential extraction procedures, the mode of occurrence of trace element in coal can be classified into five fractions (exchangeable, carbon bound, Fe-Mn oxides bound, organic matter bound, and residual) or six fractions (water-leachable, ion-exchangeable, organic-bound, carbonate-bound, silicate-bound, and sulfide-bound) [38, 43]. Because coal gangue is similar to coal, the sequential extraction method was also preformed to determine the modes of occurrence of trace elements in coal gangue [12, 38, 43]. For example, the six-step sequential extraction procedure was adopted by Zhou to study the modes of occurrence of trace element in coal gangue [15]. Also, the six-step sequential extraction procedure was modified to study the modes of occurrence of Hg in four coal gangues in our previous work [12]. Generally, the sequential extraction method is an effective way to identify the modes of occurrence of trace element in coal gangue.

2.3. Mode of occurrence of mercury

Mercury is a toxic element that is hazardous to the environment and all living organisms, including human beings [45]. Six sequential extraction procedures were performed to identify the modes of occurrence of mercury in the four coal gangues (ED, GD, PL, and TX) in our previous work and the result is shown in **Table 2** [12].

It can be seen from **Table 2** that the sulfide-bound mercury is the dominant form of mercury in the four coal gangues; the levels of sulfide-bound mercury are 74.22, 51.49, 78.88, and 74.70% for ED, GD, PL, and TX, respectively; this result is similar to the results of other reports on coal [43, 46]. The result also shows that the second most abundant mercury form is the silicate-bound mercury, accounting for 10.32, 20.81, 18.55, and 21.96% of the total

Sample	Ion-exchangeable	Carbonate bound	Iron-manganese oxide	Organic bound	Sulfide bound	Silicate bound	Residue
ED	0.91	0.38	Undetected	2.41	74.22	10.32	13.85
GD	0.67	1.76	11.48	0.09	51.49	20.81	14.33
PL	0.98	0.08	6.28	0.27	78.88	18.55	9.49
TX	0.27	0.77	Undetected	3.40	74.70	21.96	8.39

Table 2. Distribution of Hg in the sequential extractions of the coal gangue samples (%).

mercury in the coal gangues of ED, GD, PL, and TX, respectively. This result is close to the amount of silicate-bound mercury in the coal considered in other literature reports [47]. Iron-manganese oxide-bound mercury accounts for 11.48 and 6.28% in GD and PL coal gangue, whereas it is undetected in ED and TX coal gangue. Moreover, the content of ion-exchangeable mercury for ED, GD, PL, and TX are calculated to be 0.91, 0.67, 0.98, and 0.27%, respectively, being almost negligible at less than 1%. Similar to ion-exchangeable mercury, carbonate-bound mercury is also nearly negligible and accounts for 0.38–1.76% of the total mercury in coal gangue; this result is consistent with the other reports on coals [43, 46]. Organic-bound mercury accounts for approximately 2–4% in ED and TX and less than 0.3% in GD and PL, whereas the content of organic-bound mercury almost exceeded 40% in coal [43]. This result indicates that a small portion of organic-bound mercury exists in coal gangue because of the high content of minerals and low content of organic matter in coal gangue. However, a portion of mercury remains in the residue, which is possibly insoluble mercury, such as $HgSO_4$ or HgS [48]. As a result, the modes of occurrence of mercury in the four coal gangues is generally in the order of sulfide-bound mercury > silicate-bound mercury > iron-manganese oxide-bound mercury > organic-bound mercury > carbonate-bound mercury ≈ ion-exchangeable mercury [12].

2.4. Mode of occurrence of arsenic

Arsenic is considered as an environmentally sensitive element [49]. Extensive studies have demonstrated that arsenic is mainly associated with pyrite [5, 9, 50, 51]. The sequential extraction procedures of arsenic in Huainan coal gangue show that a major portion is bound to the sulfide fraction, followed by the silicate, carbonate, and organic fractions, which is consistent with the previous hypothesis that arsenic is predominately associated with pyrite [38]. In addition, the sequential extraction procedures were also performed to identify the modes of occurrence of arsenic in three coal gangues in Shanxi province by Cao et al. [52]. The modes of occurrence of arsenic could be characterized as ion-exchangeable, carbonate-bound, iron-manganese oxide-bound, sulfide-bound, organic-bound, and residual forms. The experimental results show that the sulfide-bound arsenic is dominant form, accounting for 56.71–79.36% of the arsenic in the coal gangues. The residual arsenic is the second most abundant form, accounting for 11.47–18.02% of the total arsenic. The organic-bound arsenic is the third most abundant form in the coal gangues. However, the ion-exchangeable arsenic, the carbonate-bound arsenic, and iron-manganese oxide-bound arsenic in the coal gangues are all less than 2% [52].

2.5. Mode of occurrence of other trace elements

The modes of occurrence of other trace elements in coal gangue were also identified by the six-step sequential extraction procedures [38, 53]. The result shows that the trace elements of Co, Cr, Cu, Mn, Ni, Se, Sn, V, and Zn in Huainan coal gangue are predominantly existed in silicate-bound, sulfide-bound, and carbonate-bound forms while a small portion of them was found to be associated with the organic-bound, ion-exchangeable, and water-leachable form [38]. The positive correlation between the trace element concentration in Huainan coal gangue with ash yield, aluminum, calcium, and iron-sulfur has been found, indicating that the trace elements in Huainan coal gangue are mainly associated with sulfide minerals, which could release out from coal gangue easily and can disperse into the environment [38]. Meanwhile, the results show that the trace elements of Be, V, Cr, Co, Ni, Cu, Sr, Mo, Cd, Sb, Ba, Pb, Th, U, and Se in coal gangue from Pingdingshan Mine Area mainly exist in the residual form, then in carbonate-bound form, iron and manganese oxides form, organic matter and sulfide form, and the percentage of the exchangeable form is relatively low [53]. For one mode of occurrence of trace element, the potential hazardous to environment are different with other modes of occurrence of trace element [53].

Overall, the sulfide-bound trace elements are dominant form in coal gangue, which should be ascribed to the large amount of pyrite in coal gangue. Silicate-bound trace elements also are one of the main forms, which is due to the silicate mineral in coal gangue.

3. Leaching character of trace elements from coal gangue

3.1. Leaching method

Extensive studies on the leaching behavior of trace elements from coal, coal fly ash, and bottom ash have been conducted and it proves that leaching is one of the primary pathways for trace elements entering into the ecosystem [9, 54, 55]. Therefore, some studies about the leaching behavior of trace elements from coal gangue, especially from coal gangue piles, have been done by some authors [37, 56]. The leaching behavior of the trace elements from coal gangue indicated that leaching is a very complex process, which might be affected by many factors. Generally, the pH of the leaching solutions plays an important role on the leaching behavior of trace elements from coal gangue [56]. Also, leaching time is another important impact factor on the leaching behavior of trace elements from coal gangue [57–59]. Meanwhile, the concentration of the trace elements in coal gangue can influence the leaching behavior [37].

3.2. Leaching behavior of trace elements from coal gangue under column leaching experiments

The study by Yang et al. was conducted to focus on investigations on the leaching behavior of trace elements from coal gangue in an open-cast coal mine, Inner Mongolia, China [37]. Four comparative column leaching experiments were carried out to investigate the influences of leaching time, the pH values of solution, and the sample amount on the leaching behavior of trace elements from coal gangue. Based on the results of comparative column leaching experiments of coal gangue, the environmental and ecological risks caused by the trace elements of the resulting leachates from coal gangue piles were evaluated in detail by some semi-quantitative methods. In the study, the leached concentrations, leached amount, leachability, maximum leached amount, and maximum leachability of trace elements were used to illustrate the leaching behavior of the trace elements from coal gangue piles. The leaching behavior of the 12 trace elements in coal gangue piles including As, V, Cr, Co, Ni, Cu, Zn, Se, Cd, Sn, Pb, and Hg was studied [37].

The result showed that leaching time and sample amount presented significantly influence on leaching behavior of trace elements such as leached concentrations, leached amounts, and leachability. This result is consistent with the results of other literatures [52, 57, 58, 60]. For example, column leaching experiments were performed on the samples of feed coal, fly ash, and bottom ashes from the Shizuishan coal-fired power plant to investigate the leaching behavior of the 11 potentially hazardous elements and the results show that the leaching behavior of the elements from the solid samples is mainly influenced by leaching time and pH value of solution [57]. Also, different acid solutions were used to investigate the leaching behavior of bromine in coal and the results show that leaching time and pH value played an important role in controlling the leaching behavior of bromine in coal [58]. Overall, the leaching behavior of the trace elements was mainly affected by the pH values of the leaching solutions [57–60]. However, the pH values of the leaching solution showed a weak effect on the leaching behavior of the trace elements in this study. The leaching result at different sample amount indicated that the sample amount showed little effect on the change trends of the concentration of the trace element with time while it showed a remarkable effect on the concentrations of the elements in the resulting leachates, indicating that the sample amount possibly had an effect on the leaching behavior of some hazardous trace elements [37].

3.3. Leaching behavior of trace elements from coal gangue under simulation leaching conditions

A leaching test of trace elements (Cr, Mn, Cu, Zn, Pb, and Cd) in the coal gangues from the Jiulong Coal Mine was carried out in the laboratory under simulation leaching conditions by Zhang and Ouyang [56]. The influences of pH values on the leaching behavior from the coal gangue samples were studied through static leaching experiments. At the same time, the trace elements of raining water around the coal gangue dumps were also investigated, which can give valuable information about the migration mechanism and influence factors of trace element during leaching process of coal gangues. The results showed that the concentration of trace elements dissolved out during the leaching, and the release rate of trace elements from coal gangue was not only related to the pH value of solution but also related to the trace element concentrations in coal gangues. The concentrations of trace elements in the filter water and rainwater collected around the coal gangue dumps were both higher than the standards limit of the third-class surface water according to the environmental quality standards.

The leaching rates of trace elements are different with different pH value of solutions. The leaching rates of Zn and Cd can reach the maximum value in solutions of pH = 7 under the static leaching experiment. The leaching rates of Cr, Mn, Ni, Cu, and Pb reach a minimum value in solutions of pH = 5 under the static leaching experiment. Among all the trace elements, Mn presents the minimum release rate. The concentration of trace elements in rainwater sample is in the order of Zn > Ni > Cr > Cu > Mn > Pb > Cd. The concentration of trace elements in leachate is in the order of Zn > Ni > Cu > Cr > Mn > Pb > Cd, which shows a similar tendency with that in rainwater sample except Cu and Cr. It should be noted that the concentrations of Mn, Cu, Zn, Cr, Cd, and Pb are all higher than the standards limit of the third-class surface water of the environmental quality standards, indicating that the leaching of trace elements from coal gangue can cause potential environment problems [56].

3.4. Leaching characteristics of trace elements in fresh, weathering, and filling coal gangue

The fresh, weathering, and filling coal gangue samples, as well as coal slime were collected from Xinzhuangzi Coal Mine in Huainan area [61]. The batch leaching experiments of trace elements (Cd, Cr, Pb, and Mn) from all the samples were carried out under different leaching solution and different leaching time with different particle size of the sample. The concentration of trace elements (Cd, Cr, Pb, and Mn) in different leaching solutions was measured by graphite furnace atomic absorption spectrometry. The leaching concentration and total leaching rate of trace elements were combined to analyze the leaching characteristics of the trace element in the samples. Meanwhile, the environmental effects of trace elements were assessed according to the environmental quality standard of underground water. The results show that the leaching efficiency is enhanced with the decline of particle size of sample and pH of leaching solutions and leaching amount of trace elements is increased as time prolongs. The total leaching rate of trace elements in coal slime is higher than that of other coal gangues. The total leaching rate of trace elements in weathering coal gangue is less than other samples. Cd in all the samples presents the maximum leaching performance among all the trace element studied, indicating that Cd is the most hazardous trace element under the same leaching conditions of sample type, particle size of sample, and leaching solution. Mn shows the minimum leaching performance among all the trace elements and almost has no risk to the environment. The leaching performance of the four trace elements are in the order of Cd > Pb > Cr > Mn [61].

In summary, leaching behavior of trace elements from coal gangue is affected by many factors. Leaching is the main pathway for trace elements in coal gangue into the water environment. More attention should be paid on the toxic trace elements in water environment near coal gangue piles.

4. Release character of trace element during combustion of coal gangue

4.1. Trace element partitioning behavior of coal gangue-fired CFB plant

Combustion of coal gangue in power generation is a promising method of energy recovery of coal gangue. However, the prevention of toxic trace metal emission is a significant concern during coal gangue combustion.

The trace element partitioning behavior of the Pingshuo coal gangue-fired power plant in Shanxi province of China has been investigated [15]. The feed fuel, bottom ash, and fly ash samples were collected from the coal gangue-fired power plant. The experimental analysis and the thermodynamic equilibrium calculation were conducted to study the trace element partitioning behavior during coal gangue combustion. The result of analysis shows that the trace elements can be divided into three groups, including highly volatile elements, semi-volatile elements, and non-volatile elements. The highly volatile elements including Hg, As, Be, and Cd, may be emitted into the atmosphere through the gas phase or as fine particles in the flue gas. The semi-volatile elements contain Pb, Co, Zn, Cu, and Ni, which may be enriched in the fly ash relative to the bottom ash. The non-volatile elements contain Cr and Mn, which are relatively enriched in the bottom ash. The result of thermodynamic equilibrium calculation shows that the existence of chlorine may increase the volatility of several trace elements and the presence of mineral phases such as aluminosilicates can decrease the volatility of elements by chemical immobilization effect. The air/fuel ratio and the mode of occurrence of trace elements may also largely influence the release behavior of trace elements [15].

A study on some toxic trace elements behaviors of a 330 MW coal gangue CFB power plant in Huainan of China was performed by Zhou et al. [1]. The feed fuel, bottom ash, fly ash, and flue gas samples were all collected. The partitioning behavior of toxic elements in the power plants was analyzed systemically. The result shows that the toxic elements can be classified into three groups. The first group includes the elements of As, Cd, Cu, Pb, Se, and Sn, which have high volatile tendencies with the volatilization ratio more than 20% and are mainly enriched in fly ash. The second group includes the elements of Co, Cr, Mn, and V, which have low volatilization rate and are equally distributed between bottom ash and fly ash. The third group includes the element of Bi, Ni, and Zn, which have moderate volatile tendencies. The modes of occurrence of trace elements have significant effect on the transformation behaviors of toxic elements during coal gangue combustion [1].

4.2. Trace element behavior during spontaneous combustion of coal gangue

When the rate of heat generated by the oxidation of organic matter or pyrite in coal gangue exceeds the rate of heat dissipation, the spontaneous combustion of coal gangue can be occurred [62]. Because the amount of coal gangue is very high and coal gangue spontaneous combustion is widespread in Yangquan of Shanxi Province, the coal gangue dumps in this area were investigated to study the trace element behavior during coal gangue spontaneous combustion [3].

All kinds of coal gangue samples including bulk gangue, specific lithologies, fired coal gangue and condensate products from gas vents were fully characterized regarding the mineralogy, chemistry, and leaching potential [3]. The result showed that the temperature during the spontaneous combustion of coal gangue could reach to 1200°C, which was also verified by laboratory calcination tests [3]. During spontaneous combustion of coal gangue, some elements (C, Cl, F, As, Cd, Hg, Pb, Sn, Ge, and Se) may be released into the atmosphere, which well agreed with Zhao [4]. However, the gaseous emissions of As, S, N, Hg, and Se can be partially trapped by the important condensation processes. Therefore, the condensate mineralizations enriched in Se, As, and other trace elements in the gas vent [3]. It should be pointed out that a quick cover of soil on coal gangue can help to avoid oxidation of the organic material or the pyrite and thus reduce the risk of self-ignition of coal gangue [3].

4.3. Transformation behavior of trace elements during coal gangue combustion

It is of great interest to study the trace element behaviors concerning coal gangue combustion in order to understand their fate during coal gangue combustion. Several factors, such as volatilization tendency of trace elements, modes of occurrence of trace elements, and concentrations of trace elements in coal gangue may influence the trace elements behavior during coal gangue combustion [63–66]. Also, the combustion condition, especially temperature is a key factor to influence the transformation behavior of trace elements.

The transformation behavior of trace elements during combustion of coal gangue under different temperatures had been studied by Zhou et al. [14]. The coal gangue samples were collected from the Xinzhuangzi Mine, China. Eight combustion temperatures (500, 600, 700, 800, 900, 1000, 1100, and 1200°C) were selected to study the trace elements behavior during coal gangue combustion. The coal gangue sample was placed in a muffle roaster at determined temperature and sustained a time of 30 min. Then, the coal gangue samples were taken out from muffle roaster and cooled down to the room temperature in a dryer. The content of trace elements (Ni, Cu, Zn, Cd, Sn, Pb, As, Co, Cr, and V) in coal gangue and combustion ash was determined by inductively coupled plasma mass spectrometry (ICP-MS). The results show that the trace element behavior during coal gangue combustion mainly relies on the combustion temperature. The volatilize ratios of trace elements increase with increasing combustion temperature. It indicates that higher temperature could lead to a relatively thorough coal gangue thermal decomposing, and the trace elements originally existed in coal gangue could easily release out. Moreover, 10 trace elements studied in the current study could be divided into two groups according to their volatilize ratios. The first group contains Ni, Cu, Zn, Cd, Sn, Pb, and As, which show a high volatilize tendencies. The second group is represented by Co, Cr, and V, which were relative non-volatile. It is reported that the volatility of trace elements during coal combustion mainly relies on the modes of occurrence and the concentrations in coals as well as the combustion parameters [64–67]. During coal gangue combustion, the combustion temperature and the modes of occurrence of trace elements in coal gangue are the main factors. However, the content of trace elements in coal gangue was not the key factor to influence its transformation behavior. For example, the Cd content in the coal gangue was quite low, but its volatilize ratio was quite high.

In a word, different trace elements presented different transformation behaviors because of the different thermal stability and the different modes of occurrence of trace elements in coal gangue.

5. Environmental implication of trace element in coal gangue

5.1. Potential environmental effects of trace elements in coal gangue of an open-cast coal mine area

The environmental effects of trace elements from coal gangue piles of an open-cast coal mine area, Inner Mongolia, China have been investigated by Yang et al. Twelve trace elements, including As, V, Cr, Co, Ni, Cu, Zn, Se, Cd, Sn, Pb, and Hg, were selected for investigation to assess their environmental and ecological hazards levels in coal gangue piles [37].

The enrichment factor was an important parameter to evaluate the pollution level of an element [38, 68]. Therefore, the enrichment factor values of elements in coal gangue were applied to assess the potential environmental effect of the 12 trace elements [37]. The results show that the enrichment factor values for elements of As and Se were 4.86 and 7.41, respectively, indicating that they can cause a high pollution level due to their high concentration in the coal gangue.

It is reported that "Maximum leached amount" (Lam) can act as an important indicator for trace element environmental risk assessment because it can provide valuable information of the maximum transformation ability of element in environment [57]. Thus, the value of Lam is used to evaluate the environmental risk caused by trace element in coal gangue. Meanwhile, the concentration limits of the trace elements in groundwater and soil were also investigated in this research. The results show that the Lam values of the 12 trace elements are all less than the corresponding concentrations limits of the elements in the environmental quality standard for soils as well as the limits of contaminants in foods. It indicated that the 12 trace elements in the coal gangue almost had no potential risk to soils and vegetation. However, since the Lam values of some trace elements are higher than the corresponding concentration in the quality standard for groundwater, the trace elements could possibly have some potential impacts on the groundwater. Meanwhile, the evaluation result of biological toxicity showed that the 12 trace elements from coal gangue could produce moderate to high level of ecological risk to the environment [37].

5.2. Environmental implications of trace elements associated with coal gangue

In order to evaluate the potential environmental implications of the trace elements associated with coal gangue, the mobility behavior of 10 trace elements (As, Co, Cr, Cu, Mn, Ni, Se, Sn, V, and Zn) in coal gangue from the Huainan Coalfield was studied by Zhou et al. [38]. The sequential extraction was also used to analyze the chemical properties of trace elements in coal gangue because the toxicity of trace elements mainly depends on its chemical properties. The result showed that the chemical properties of trace elements in the coal gangue are mainly as silicate-bound, sulfide-bound and carbonate-bound form. The positive correlation of the trace element concentration with ash yield or iron-sulfur material indicated that the trace elements of As, Co, Cu, Ni, Se, and Zn in coal gangue are basically bounded with sulfide minerals in coal gangue. Therefore, they could release out and disperse into the environment easily during the natural weathering of coal gangue, leading to potential hazardous to environment.

To evaluate the biological toxicity of trace elements of Cr, Cu, Zn, Ni, and As in coal gangue, international sediment quality guidelines calculations such as effects range low (ERL) and effects range median (ERM) were adopted [69, 70]. The result showed that the trace elements of Zn, Cr, and Cu are below ERLs, indicating unobvious adverse biological effects. The trace elements of Ni and As are in the middle range, indicating occasional adverse biological effects. None of the trace elements in the samples are higher than ERMs value, suggesting limited adverse biological effects [38]. The result of the risk assessment code reveals that the trace elements of Mn, Cr, Se, Ni, Zn, As, and Cu can pose serious environmental risks to the ecosystem while Co, Sn, and V pose no risk or low risk [38].

5.3. Environmental characterization of burnt coal gangue banks

The coal gangue dumps in Yangquan of Shanxi Province, China were selected to investigate the environmental characteristics of burnt coal gangue banks [3]. The fresh coal gangue, weathered coal gangue, and the gas vents condensates were sampled and studied by leaching technique to evaluate their environment impacts. The result showed that the trace elements in the fresh coal gangue presented relatively low leaching potential. However, the weathered coal gangue and the gas vents condensates could produce acidic leachates, leading to relatively high leachable levels of trace elements and thus giving rise to environmental problems [3].

In summary, trace elements in coal gangue could release out and produce environmental implication in various degrees, depending on the type of trace elements. The potential environmental effects should not be neglected and some measures to eliminate them should be taken.

6. Conclusion

Trace elements in coal gangue could be associated with the organic matter or inorganic matter in coal gangue. The sequential extraction method is an effective way to identify the modes of occurrence of trace element in coal gangue. The modes of occurrence of trace elements are different with coal gangue type. The sample amount, pH values of solution or trace element concentrations affect the leaching behavior of trace elements. The leaching characteristics of trace elements in fresh, weathering, and filling coal gangue are different. During coal gangue combustion, the trace elements can be divided into three groups, including highly volatile elements, semi-volatile elements, and non-volatile elements. The modes of occurrence of trace elements have significant effect on the behaviors of trace elements. The trace elements could possibly have some potential impacts on the groundwater and could produce moderate to high level of ecological risk to the environment. Compared with fresh coal gangue, the weathered coal gangue and the gas vents condensates presented relatively high leaching potential and could give rise to environmental problems.

Generally, the study of trace elements in coal gangue is still limited. Most of experimental method adopted were traditional methods in geology area and some modern technique should be taken, such as online determining method. Also, the study about the removal of toxic trace elements is scare and should be done at present; and the controlling technology of toxic trace elements during coal gangue utilization should be developed to alleviate the adverse impact made to the environment.

Acknowledgements

The authors gratefully acknowledge the financial support from the Natural Science Foundation of China (41372350).

[22] Font O, Querol X, Izquierdo M, Alvarez E, Moreno N, Diez S, Alvarez-Rodriguez R, Clemente-Jul C, Coca P, Garcia-Pena F. Partitioning of elements in a entrained flow IGCC plant: Influence of selected operational conditions. Fuel. 2010;**89**:3250-3261. DOI: 10.1016/j.fuel.2010.03.044

[23] Kukier U, Ishak CF, Sumner ME, Miller WP. Composition and element solubility of magnetic and non-magnetic fly ash fractions. Environmental Pollution. 2003;**123**:255-266. DOI: 10.1016/S0269-7491(02)00376-7

[24] Raeva AA, Dongari N, Artemyeva AA, Kozliak EI, Pierce DT, Seames WS. Experimental simulation of trace element evolution from the excluded mineral fraction during coal combustion using GFAAS and TGA-DSC. Fuel. 2014;**124**:28-40. DOI: 10.1016/j.fuel.2014.01.078

[25] MH X, Yan R, Zheng CG, Qiao Y, Han J, Sheng CD. Status of trace element emission in a coal combustion process: A review. Fuel Processing Technology. 2004;**85**:215-237. DOI: 10.1016/S0378-3820(03)00174-7

[26] Bhangare RC, Ajmal PY, Sahu SK, Pandit GG, Puranik VD. Distribution of trace elements in coal and combustion residues from five thermal power plants in India. International Journal of Coal Geology. 2011;**86**:349-356. DOI: 10.1016/j.coal.2011.03.008

[27] Swanson SM, Engle MA, Ruppert LF, Affolter RH, Jones KB. Partitioning of selected trace elements in coal combustion products from two coal-burning power plants in the United States. International Journal of Coal Geology. 2013;**113**:116-126. DOI: 10.1016/j.coal.2012.08.010

[28] Diaz-Somoano M, Martinez-Tarazona MR. Trace element evaporation during coal gasification based on a thermodynamic equilibrium calculation approach. Fuel. 2003;**82**:137-145. DOI: 10.1016/S0016-2361(02)00251-X

[29] Jano-Ito MA, Reed GP, Milian M. Comparison of thermodynamic equilibrium predictions on trace element speciation in oxy-fuel and conventional coal combustion power plants. Energy & Fuel. 2014;**28**:4666-4683. DOI: 10.1021/ef5005607

[30] Lundholm K, Nordin A, Backman R. Trace element speciation in combustion processes-review and compilations of thermodynamic data. Fuel Processing Technology. 2007;**88**:1061-1070. DOI: 10.1016/j.fuproc.2007.06.032

[31] Miller BB, Kandiyoti R, Dugwell DR. Trace element behavior during co-combustion of sewage sludge with polish coal. Energy & Fuel. 2004;**18**:1093-1103. DOI: 10.1021/ef0400181

[32] Thompson D, Argent BB. Thermodynamic equilibrium study of trace element mobilisation under pulverised fuel combustion conditions. Fuel. 2002;**81**:345-361. DOI: 10.1016/S0016-2361(01)00145-4

[33] Yang JP, Zhao YC, Zhang JY, Zheng CG. Removal of elemental mercury from flue gas by recyclable $CuCl_2$ modified magnetospheres catalyst from fly ash. Part 3. Regeneration performance in realistic flue gas atmosphere. Fuel. 2016;**173**:1-7. DOI: 10.1016/j.fuel.2015.12.077

[34] Lopez-Anton MA, Diaz-Somoano M, Spears DA, Martinez-Tarazona MR. Arsenic and selenium capture by fly ashes at low temperature. Environmental Science & Technology. 2006;**40**:3947-3951. DOI: 10.1021/es0522027

[35] Zhou C, Liu G, Wu S, Lam PKS. The environmental characteristics of usage of coal gangue in bricking-making: A case study at Huainan, China. Chemosphere. 2014;**95**:274-280. DOI: 10.1016/j.chemosphere.2013.09.004

[36] Zhou CC, Liu GJ, ZY X, Sun H, Lam PKS. The retention mechanism, transformation behavior and environmental implication of trace element during co-combustion coal gangue with soybean stalk. Fuel. 2017;**189**:32-38. DOI: 10.1016/j.fuel.2016.10.093

[37] Yang L, Song JF, Bai X, Song B, Wang RD, Zhou TH, Jia JL, Pu HX. Leaching behavior and potential environmental effects of trace elements in coal gangue of an open-cast coal mine area, Inner Mongolia, China. Minerals. 2016;**6**:50. DOI: 10.3390/min6020050

[38] Zhou CC, Liu GJ, Wu D, Fang T, Wang RW, Fan X. Mobility behavior and environmental implications of trace elements associated with coal gangue: A case study at the Huainan coalfield in China. Chemosphere. 2014;**95**:193-199. DOI: 10.1016/j.chemosphere.2013.08.065

[39] Vassilev SV, Menendez R, Diaz-Sornoano M, Martinez-Tarazona MR. Phase-mineral and chemical composition of coal fly ashes as a basis for their multicomponent utilization. 2. Characterization of ceramic cenosphere and salt concentrates. Fuel. 2004;**83**:585-603. DOI: 10.1016/j.fuel.2004.11.021

[40] Vassileva CG, Vassilev SV. Behaviour of inorganic matter during heating of Bulgarian coals-2. Subbituminous and bituminous coals. Fuel Processing Technology. 2006;**87**:1095-1116. DOI: 10.1016/j.fuproc.2006.08.006

[41] Ketris MP, Yudovich YE. Estimations of clarkes for carbonaceous biolithes: World average for trace element contents in black shales and coals. International Journal of Coal Geology. 2009;**78**:135-148. DOI: 10.1016/j.coal.2009.01.002

[42] Tessier A, Campbell P, Blsson M. Sequential extraction procedure for the speciation particulate trace metals. Analytical Chemistry. 1979;**51**:844-851. DOI: 10.1021/ac50043a017

[43] Zheng L, Liu G, Qi C, Zhang Y, Wong M. The use of sequential extraction to determine the distribution and modes of occurrence of mercury in Permian Huaibei coal, Anhui Province, China. International Journal of Coal Geology. 2008;**73**:139-155. DOI: 10.1016/j.coal.2007.04.001

[44] Biester H, Scholz C. Determination of mercury binding forms in contaminated soils: Mercury pyrolysis versus sequential extractions. Environmental Science & Technology. 1997;**31**:233-239. DOI: 10.1021/es960369h

[45] Sundseth K, Pacyna JM, Pacyna EG, Munthe J, Belhaj M, Astrom S. Economic benefits from decreased mercury emissions: Projections for 2020. Journal of Cleaner Production. 2010;**18**:386-394. DOI: 10.1016 / j.jclepro. 2009.10.017

[46] Feng X, Hong Y. Modes of occurrence of mercury in coals from Guizhou, People's Republic of China. Fuel. 1999;**78**:1181-1188. DOI: 10.1016/S0016-2361(99)00077-0

[47] Zhang J, Ren D, Zhu Y, Chou CL, Zeng R, Zheng B. Mineral matter and potentially hazardous trace elements in coals from Qianxi fault depression area in southwestern Guizhou, China. International Journal of Coal Geology. 2004;**57**:49-61. DOI: 10.1016/j.coal.2003.07.001

[48] Guo S, Yang J, Liu Z. Characterization of hg in coals by temperature programmed decomposition atomic fluorescence spectroscopy and acid-leaching technique. Energy & Fuels. 2012;**26**:3388-3392. DOI: 10.1021/ef201598d

[49] Swaine DJ. Why trace elements are important. Fuel Processing Technology. 2000;**65**:21-33. DOI: 10.1016/S0378-3820(99)00073-9

[50] Huggins FE, Huffman GP. Modes of occurrence of trace elements in coal from XAFS spectroscopy. International Journal of Coal Geology. 1996;**32**:31-53. DOI: 10.1016/S0166-5162(96)00029-8

[51] Ward CR. Analysis and significance of mineral matter in coal seams. International Journal of Coal Geology. 2002;**50**:135-168. DOI: 10.1016/S0166-5162(02)00117-9

[52] Cao Y, Guo S, Zhai J. The mode of occurrence of mercury and arsenic in coal gangues. Coal Geology & Exploration. 2017;**45**:26-30. DOI: 10.3969/j.issn.1001-1986.2017.01.005 (in Chinese edition)

[53] Xu H, Guo H, Song W, Lu H, Zhang Z, Zhang G. Modes of occurrence of trace elements in coal gangue from NO.1 coal mine of Pingdingshan mine area. Environmental Science and Technology. 2016;**39**:65-69. DOI: 10.3969/j.issn.1003-6504.2016.05.013 (In Chinese edition)

[54] Praharaj T, Powell MA, Hart BR, Tripathy S. Leachability of elements from sub-bituminous coal fly ash from India. Environment International. 2002;**27**:609-615. DOI: 10.1016/S0160-4120(01)00118-0

[55] Izquierdo M, Querol X. Leaching behaviour of elements from coal combustion fly ash: An overview. International Journal of Coal Geology. 2012;**94**:54-66. DOI: 10.1016/j.coal.2011.10.006

[56] Zhang HJ, Ouyang SL. Release characteristics of heavy metals from coal gangue under simulation leaching conditions. Energy Exploration and Exploitation. 2014;**32**:413-422

[57] Wang W, Qin Y, Song D, Wang K. Column leaching of coal and its combustion residues, Shizuishan, China. International Journal of Coal Geology. 2008;**75**:81-87. DOI: 10.1016/j.coal.2008.02.004

[58] Peng B, Leaching WD. Characteristics of bromine in coal. Journal of Fuel Chemistry and Technology. 2011;**39**:647-651. DOI: 10.3969/j.issn.0253-2409.2011.09.002 (In Chinese edition)

[59] Huggins FE, Seidu LBA, Shah N, Backus J, Huffman GP, Honaker RQ. Mobility of elements in long-term leaching tests on Illinois #6 coal rejects. International Journal of Coal Geology. 2011;**94**:326-336. DOI: 10.1016/j.coal.2011.04.006

[60] Spears DA. The determination of trace element distributions in coals using sequential chemical leaching—A new approach to an old method. Fuel. 2013;**114**:31-37. DOI: 10.1016/j.fuel.2012.09.028

[61] Zheng L, Ding S, Liu C, Jiang C, Chen Y. Leaching characteristics of environmentally sensitive trace elements in different types of coal gangue. Journal of Central South University of Science and Technology. 2016;**47**:703-710. DOI: 10.11817/j.issn.1672−7207.2016.02.048 (In Chinese edition)

[62] Misra BK, Singh BD. Susceptibility to spontaneous combustion of Indian coals and lignites: An organic petrographic autopsy. International Journal of Coal Geology. 1994;**25**:265-286. DOI: 10.1016/0166-5162(94)90019-1

[63] Querol X, Fernandez-Turiel JL, Lopez-Soler A. Trace elements in coal and their behavior during combustion in a large power station. Fuel. 1995;**74**:331-343. DOI: 10.1016/0016-2361(95)93464-O

[64] Furimsky E. Characterization of trace element emissions from coal combustion by equilibrium calculations. Fuel Processing Technology. 2000;**63**:29-44. DOI: 10.1016/S0378-3820(99)00067-3

[65] Liu GJ, Zhang HY, Gao LF, Zheng LG, Peng ZC. Petrological and mineralogical characterizations and chemical composition of coal ashes from power plants in Yanzhou mining district. China. Fuel Processing Technology. 2004;**85**:1635-1646. DOI: 10.1016/j.fuproc.2003.10.028

[66] Senior CL, Bool LE III, Srinivasachar S, Pease BR, Porle K. Pilot scale study of trace element vaporization and condensation during combustion of a pulverized sub-bituminous coal. Fuel Processing Technology. 2000;**63**:149-165. DOI: 10.1016/S0378-3820(99)00094-6

[67] Guo RX, Yang JL, Liu ZY. Influence of heat treatment conditions on release of chlorine from Datong coal. Journal of Analytical and Applied Pyrolysis. 2004;**71**:179-186. DOI: 10.1016/S0165-2370(03)00086-X

[68] Wang W, Hao W, Bian Z, Lei S, Wang X, Sang S. Effect of coal mining activities on the environment of Tetraenamongolica inWuhai, Inner Mongolia, China—A geochemical perspective. International Journal of Coal Geology. 2014;**132**:94-102. DOI: 10.1016/j.coal.2014.08.006

[69] Long E, MacDonald D, Smith S, Calder F. Incidence of adverse biological effects within ranges of chemical concentrations in marine and estuarine sediments. Environmental Management. 1995;**19**:81-97. DOI: 10.1007/BF02472006

[70] Pekey H, Karakas D, Ayberk S, Tolun L, Bakoglu M. Ecological risk assessment using trace elements from surface sediments of Izmit Bay (Northeastern Marmara Sea) Turkey. Marine Pollution Bulletin. 2004;**48**:946-953. DOI: 10.1016/j.marpolbul.2003.11.023

Chapter 7

Mineralization: Evidence from Fission Track Thermochronology

Wanming Yuan and Ke Wang

Additional information is available at the end of the chapter

http://dx.doi.org/10.5772/intechopen.72277

Abstract

Ore deposits were the product of the Earth's material movement in a certain historical stage and tend to experience different forms and different degrees of change until being found, exploited and utilized. We should attach importance to conservation and changes of ore deposits besides metallogenic environment, ore deposit model and origin research. The conservation is closely related to uplifting and denudation so that to recover histories of uplifting and denudation for ore districts could reveal conservation and changes of mineral deposits. By applying fission track thermochronology, this chapter presents a research sample to discuss the issue, especially the relative technical method, and provides evidences for both deep ore prospecting and mineralizing potentiality evaluation. Meanwhile, dating mineralizing age is another frontier topic in the world. The author successfully applied fission track thermochronology to determining the mineralizing ages and epochs of the hydrothermal deposits. Steps and methods of achieving these goals are shown in detail. Geologists could take this chapter as a reference tool.

Keywords: mineralizing age, mineralizing epoch, preservation of orebody, tectonics, fission track thermochronology

1. Introduction

How to confirm mineralizing age and epoch of hydrothermal ore deposits is a frontier issue to research on mineralization in present-day world. The uncertainty of mineralization epoch highly restricts the deep research on the mineralization regularities, deposit origin, prospective prediction, and relationship between mineralization and tectonic thermal event. As a new technique method, fission track (FT) thermochronology has definite superiority for dating mineralizing age and epoch. We have a new significant attempt to apply the fission track method to the hydrothermal mineralization. Since annealing of fission tracks occurs in crystal lattice and FT stability is hardly affected by other factors except temperature, the FT thermochronology is quite suitable for recording thermal events that take place in open system or enclosed system, in which the ore deposit formed. Reconstruction of thermal event history is a good quality of the fission track method that can indicate the variety relationship between temperature and time for the thermal event [1].

2. Fission track thermochronology

Silk and Barnes [2] first recognized track fading when they observed fission fragment tracks in muscovite mica under transmission electron microscope. Price and Walker [3] found the way to circumvent the problem of the rapid fading of tracks in muscovite mica and stabilized and fixed the tracks by chemical etching. They enlarged the tracks continuously by etching until the tracks were observed under optical microscope. Price and Walker [4] also discovered fossil tracks of ^{238}U spontaneous fission in muscovite mica and invented the technique of fission track dating. Fleischer et al. [5, 6] recognized that thermal fading of tracks follows the Arrhenius equation, and later many other scientists confirmed its correctness [7].

Fission track method has been widely used, and up to now, it is still one of basic techniques for nuclear science dating. In the basic dating formula,

$$t = \frac{1}{\lambda_D} \ln\left(\frac{\lambda_D N_F}{\lambda_F N_{238}} + 1\right) \tag{1}$$

^{238}U total decay constant $\lambda_D = 1.551 \times 10^{-10} a^{-1}$. The number of ^{238}U nuclei in spontaneous fission and that of existing N_{238} (number of existing ^{238}U nuclei) can be both accurately measured. Instead, no unified value of ^{238}U spontaneous fission constant λ_F had not been obtained till now [8], that is, λ_F values obtained by various laboratories via different methods, deviate greatly from one another. Now the three λ_F values used are indicated respectively: $(7.03 \pm 0.11) \times 10^{-17} a^{-1}$; $6.9 \times 10^{-17} a^{-1}$; $8.46 \times 10^{-17} a^{-1}$. The difference between the max and min values was large as 18.4%. When these two were inserted into Eq. (1) to calculate the age (order of 106a), they showed a difference of 18% or so. As a very complicated process, measuring N_F, that is, number of ^{238}U nuclei via chemical etching led to a high degree of uncertainty [9].

(U-Th)/He geochronology is a dating method that developed rapidly in recent years. Now this method has been widely used in crustal denudation research, near-surface substance dating and low-temperature geothermal history [10–12]. One of most important works in this technique is how to accurately determine ^{238}U, ^{232}Th, ^{147}Sm and ^{4}He contents in the mineral. Of all, neutron activation analysis (NAA) for ^{238}U, ^{232}Th and ^{147}Sm is a very advanced one. Instead, helium isotope, as a gas, makes itself difficult to measure. Based on the determination of solid-state nuclear track in apatite via thermal analysis method [13–15], the same method was employed to determine ^{4}He in apatite. This trial determination of the geological age got initial results. Here we expected this could give a reference for other scholars or inspire them.

Evidence for determination of geological age of apatite was supplied by the thermal analysis method.

The basic formula for (U-Th)/He dating is

$$N_{^4He} = 8N_{^{238}U}(e^{\lambda 238t}-1) + 7N_{^{235}U}(e^{\lambda 235t}-1) + 6N_{^{232}Th}(e^{\lambda 232t}-1) + 1N_{^{147}Sm}(e^{\lambda 147t}-1) \quad (2)$$

$N_{^4He}$, $N_{^{238}U}$, $N_{^{235}U}$, $N_{^{232}Th}$ and $N_{^{147}Sm}$ denote the numbers of ^{238}U, ^{235}U, ^{232}Th and ^{147}Sm nuclei, respectively; λ_{238}, λ_{235}, λ_{232} and λ_{147} are their respective α radioactive decay constants; t denotes the geological age of (U-Th)/He [15].

After crushing, grinding and sieving of the samples, apatites and zircons were concentrated and then separated using conventional magnetic and heavy-liquid methods. The zircon grains were mounted in glass slides, heated and covered by FEP Teflon sheets. Their external prismatic surfaces were ground and polished. Etching duration was about 20–35 h with NaOH/KOH(=1: 1) eutectic etchant at 210°C. Apatite was mounted in epoxy resin on glass slides and polished to expose internal grain surface. Spontaneous tracks were revealed by 5.5 N HNO_3 for 20 s at 21°C. Thin low-uranium muscovites as external detectors were packed together with sample grain mounts and CN2 as well as CN5 uranium dosimeter glass [16] irradiated and then etched in 40% HF for 20 min at 25°C to reveal the induced fission tracks.

Track densities for both natural and induced fission track populations were measured in air at 1000× magnification. The acceptable crystals for analysis should be prismatic sections parallel to the c-crystallographic axis. The track lengths were measured up to ~100 for each sample. Based on the IUGS-recommended Zeta calibration approach [17, 18] the fission track central ages were calculated. Errors were calculated using the techniques of Green. The χ^2 test is used to detect the probability that all age grains analyzed belong to a single population of ages [19]. A probability of <5% is evidence of an asymmetric spread of single-grain ages [20].

3. Metallotectonic event

Metallotectonic event is an important side of ore deposit research. Most mineralizations are closely related to tectonic activities, especially hydrothermal mineralization. Metallotectonics generally control ore-fluid or ore-forming material migratory concentration and ore-body location. The FT thermochronology could play a significant role in this field. Stability of fission tracks is mainly correlated to temperature. Closure temperature of apatite fission tracks (AFT) is about 100°C [21]. If the apatites are subsequently heated over the closure temperature, the tracks will be annealed. Fission tracks will form again when the temperature drops under 100°C. Since the effect of geothermal gradient, the fission track of sample in the deep (>100°C) is generally annealed and the AFT age becomes zero [22].

The metallotectonic event evolution usually corresponds with mineralizing process. Almost all fission track ages are significantly lower than their host rocks since the low annealing temperature. These age grains analyzed could belong to a single population of ages, and so the samples experienced a thermal event, which resulted in complete annealing of the original tracks subsequent to the formation of their host rocks [23, 24]. The FT ages are directly related to the distances from the fault zone. The deep part with lower age near the fault zone uplifts into the present surface because of a higher uplifting rate near the fault zone than distant from it. Although the samples were collected at similar elevation, the samples far from the fault zone display higher FT ages than those near the fault zone, since there is a higher geothermal gradient near the fault zone [25, 26].

4. Mineralization age and epoch

The FT technique method could provide information about both mineralizing ages and epochs, play a role of a geological thermometer that shows a continuous variety of relationship between temperature and time and hence constrain on the thermal history of mineralizing thermal event. The FT thermochronology has been applied to recovery of the geological thermal history widely more and more at home and abroad [27, 28].

The hydrothermal mineralizing temperature was lower, such as 100–300°C, and zircon and apatite fission tracks have the retention temperature of ~250 and ~100°C, respectively. Thus, the zircon and apatite fission track ages present the thermal event just after the mineralization. However, the polyperiodic magmatic activities and mineralization superimpositions were not only a dominant factor of huge metal accumulation but also the dynamical reason of polyperiodic long-duration mineralization [20].

The Jiapigou gold belt in the Jilin Province, northeast China, has been an important gold-producing area for more than about 190 years.

There is widespread evidence for hydrothermal alteration, and some of this may be associated with the younger intrusives, which are likely related to the arc that developed during the Mesozoic as the Mongolian seaway (paleo-Teths or paleo Okhotsk ocean) closed, and final consolidation took place between Siberia and the composite Sino-Korean-North China block. Related intrusives across the Jiapigou gold belt and surrounding areas include the large Huangnihe and Wudaoliuhe-Diaoshuihu bodies. The age of the pink moyite Wudaoliuhe-Diaoshuihu pluton, situated SE of the Jiapigou gold belt (**Figure 1**), which has an EW-trending ellipsoid shape and is 20 km long by 6 km wide, is unknown. Thus, the origin of the gold mineralization in the Jiapigou gold belt cannot be fully resolved without full knowledge of the emplacement ages and thermal history of the younger intrusives. To address this, we collected a suite of samples from across the Jiapigou gold belt at outcrop and from the mines and applied zircon geochronology, trace element analysis and fission track and argon thermochronometry.

4.1. Analytical results

4.1.1. Zircon U–Pb ages

The entire dataset, some 118 ages, yielded four age groups: 157–183, 200–213, 216–248 and 2342–2674 Ma (**Figure 2**). The youngest group of zircon U–Pb ages (166–170 or 157–183 Ma) correspond to the main early Yanshanian (Jurassic) period, the two intermediate groups of

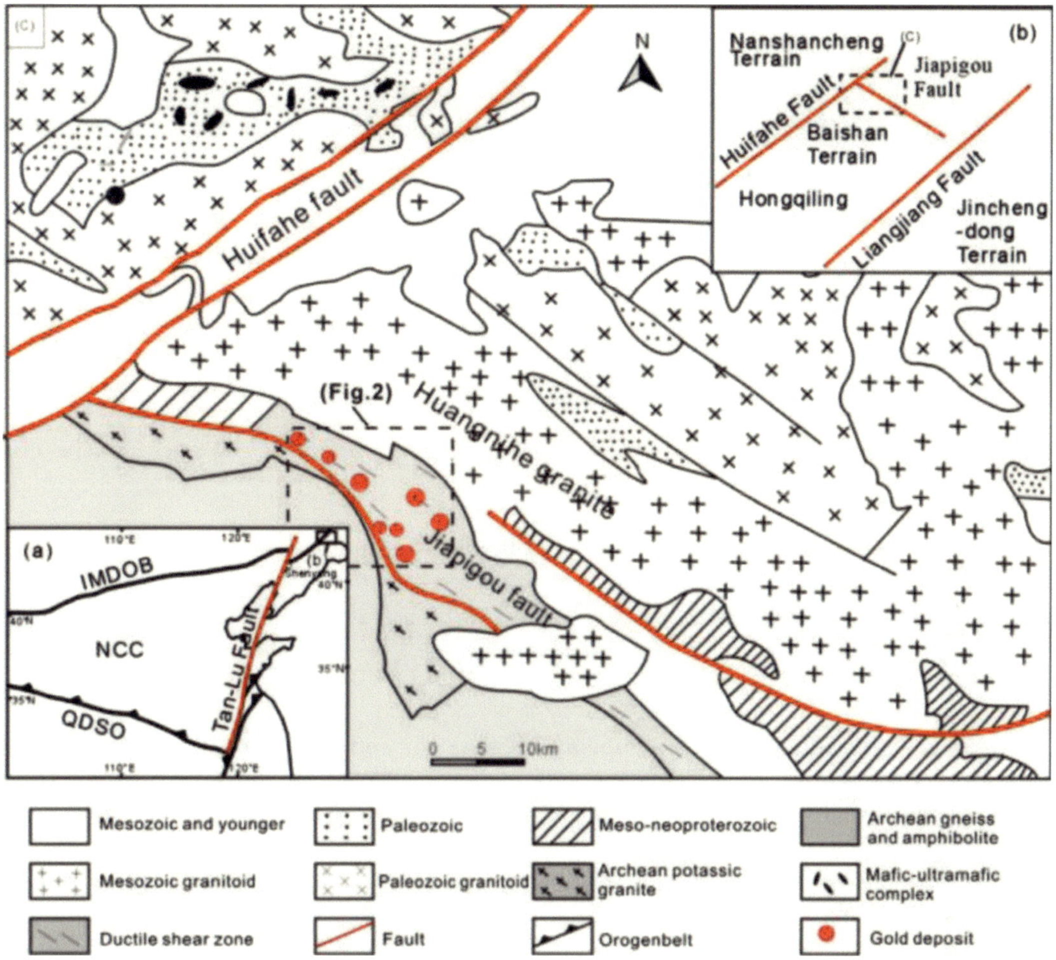

Figure 1. Geological map of the Jiapigou gold belt, eastern China. (a) Simplified map showing the location of the Jiapigou gold belt in the North China Craton (NCC) and adjacent orogenic belts, in which IMDOB is the inner Mongolian-Daxinganling orogenic belt and QDSO is Qinling-Dabie-Sulu orogen. (b) Simplified map showing the major regional faults and location of the Jiapigou gold belt. (c) General geology of the Jiapigou gold belt, showing the distribution of the major gold deposits.

zircon U–Pb ages (208 or 200–213 Ma; 228–230 or 216–248 Ma) correspond to the Indosinian (Triassic) period, and the oldest group of zircon U–Pb ages (2342–2536 or 2342–2674 Ma) correspond to the Paleoproterozoic.

4.1.2. *Zircon fission track ages*

Eleven samples were collected for ZFT analysis from different mine areas in the Jiapigou gold belt, and most of them are ore and altered rocks (**Table 1**). The results are given in **Table 1**. The central age range varies from 86 to 155 Ma. This wide age range represents various geological processes in the Jiapigou gold belt. The χ^2 test was used to examine the probability that ages

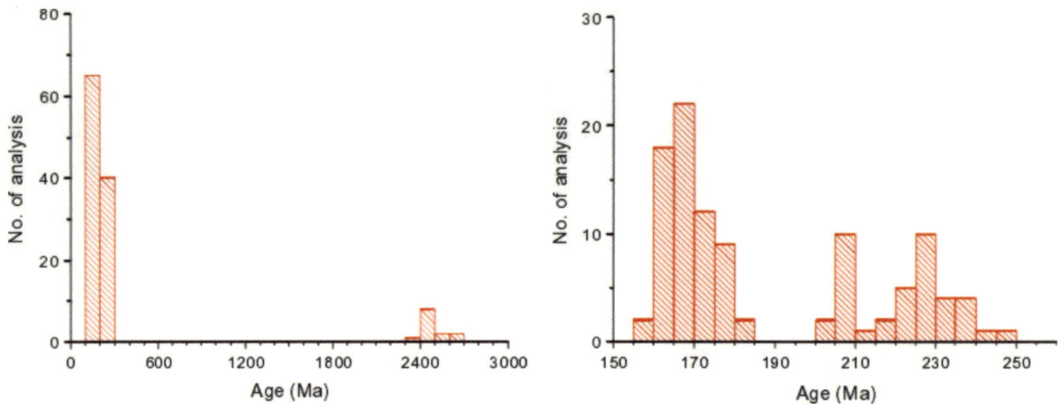

Figure 2. Histograms of the zircon U–Pb ages. Left—for all zircon grain ages; right—histogram for ages lower than 250 Ma.

obtained from zircon grains belong to a single population, using $\chi^2 > 5\%$ as statistical criterion [19]. A probability of <5% is evidence of an asymmetric spread of single-grain ages, indicating complex geneses and resources.

4.1.3. $^{40}Ar/^{39}Ar$ ages

All $^{40}Ar/^{39}Ar$ ages obtained from samples of K-feldspar and biotite as alteration minerals form age groups that are quite consistent with the groups of zircon U–Pb ages obtained from samples of magmatic rocks. Sample JP75 yielded an $^{40}Ar/^{39}Ar$ age for bioite of 228.9 ± 2.2 Ma and an $^{40}Ar/^{39}Ar$ age for K-feldspar of 217.5 ± 22.8 Ma that are similar to the zircon U–Pb age for granite of 229.8 ± 5.2 Ma. These indicate that not only K-feldspar and biotite in syenite were formed by the alteration but also that zircons in the same syenite were formed from the same metallogenic fluids that caused the formation of K-feldspar and biotite as alteration minerals. The $^{40}Ar/^{39}Ar$ plateau age of 170.3 ± 2.6 Ma for the JP88–2 K-feldspar sample denotes a younger alteration-mineralization episode that is consistent with the 170–166 Ma zircon U–Pb age group.

Among 11 samples subjected to ZFT analyses, eight samples have $P(\chi^2) > 5\%$, and three samples (BJ640-353-4-D, JP41-2, and JP100) have $P(\chi^2) < 5\%$ (**Table 1**); however, the amount of overdispersion (**Table 1**) is not large, and the pooled and central ages are the same.

The histogram of the combined ZFT data of convergent ages and decomposed divergent ages (**Figure 3**) shows three groups of ages: 86–106, 120–138 and 153–155 Ma. The number of samples increases from the highest age group to the lowest age group, which may mean enhancement of thermal event intensity from the Indosinian (Triassic) to Yanshan (Jurassic) epochs.

The bulk of the zircon U–Pb data and biotite $^{40}Ar/^{39}Ar$ age and K-feldspar $^{40}Ar/^{39}Ar$ ages obtained in this study relate to periods of alteration-mineralization in the Jiapigou gold belt. The zircon U–Pb ages indicate three periods of gold mineralization, that is, 230.0–228.9, 208.0 and 170.0–166.0 Ma. The biotite and K-feldspar $^{40}Ar/^{39}Ar$ ages of 228.9 ± 2.2 to 217.5 ± 22.8 Ma

http://dx.doi.org/10.5772/intechopen.72277

Sample	Grains (N)	ρ_s (10^5/cm^2) (Ns)	ρ_i (10^5/cm^2) (Ni)	ρ_d (10^5/cm^2) (Nd)	P(χ^2) (%)	Central age (Ma)	Pooled age (Ma)	U (ppm)
BJ640-353-4-D	16	161.621 (4655)	39.546 (1139)	6.069 (4630)	0.1	104 ± 7	105 ± 6	232.8
EDG1050-Tong-6-D	11	144.520 (3168)	30.063 (659)	6.136 (4630)	50.4	125 ± 8	125 ± 8	178.4
JL285-Zhu-13-D	9	115.404 (1384)	31.770 (381)	6.337 (4630)	13.6	98 ± 8	98 ± 7	184.1
JP100	22	167.633 (7234)	37.540 (1620)	6.605 (4630)	0.9	124 ± 8	125 ± 7	212
JP41-2	24	132.956 (7515)	33.880 (1915)	7.074 (4630)	4.5	117 ± 7	118±	175
JP54-01	23	100.335 (4892)	36.077 (1759)	7.275 (46.300)	90.8	86 ± 5	86 ± 5	182.4
JP78-2	7	164.513 (2167)	32.645 (430)	7.309 (4630)	80.6	155 ± 11	155 ± 11	166.5
JP79-2	17	140.135 (5455)	33.987 (1323)	7.175 (4630)	49	125 ± 7	125 ± 7	172.1
JXXT270-01-02	23	143.311 (6283)	39.528 (1733)	6.907 (4630)	6.5	106 ± 6	106 ± 6	206.8
JXXT270-02-04	18	161.681 (6103)	30.545 (1153)	6.852 (4630)	22.9	153 ± 9	153 ± 9	165.6
SDC740-115(100)-1D	21	141.684 (4039)	41.114 (1172)	6.789 (4630)	63.2	99 ± 6	99 ± 6	219.8

Notes: N, number of grains counted; ρ_d, induced fission track density in muscovite external detectors over SRM; Nd, total number of fission tracks counted in ρ_d; ρ_i, induced fission track density in the muscovite external detector for crystals analyzed; Ni, total number of fission tracks counted in ρ_i; ρ_s, spontaneous fission track density on the internal surfaces of apatite crystals analyzed; Ns, total number of fission tracks counted in ρ_s; P(χ^2), chi-squared probability that all single-crystal ages represent a single population of ages.

Table 1. Results of zircon fission track analyses.

and 197.9 ± 25.1 Ma correspond to the 230.0–228.9 Ma and 208.0 Ma zircon U–Pb ages of the two earlier periods of gold mineralization, respectively. The lower $^{40}Ar/^{39}Ar$ plateau age of 170.3 ± 2.6 Ma for JP88-2 K-feldspar sample is consistent with the 170.0–166.0 Ma zircon U–Pb age of the youngest period of gold mineralization.

Mineralized felsite and K-altered syenite gave the same zircon U–Pb ages of 228.9 ± 1.6 and 229.8 ± 2.0 Ma, respectively. A diabase dyke coexisting with auriferous quartz veins yielded a zircon U–Pb age of 230 ± 3.9 Ma. The age range, 228.9–230.0 Ma, could represent a period of gold metallogenesis because the samples are closely related to the gold mineralization in the Jiapiguo belt. Two other groups of zircon U–Pb ages—208 and 170–166 Ma obtained from hydrothermal zircons in a diabase dyke and granite bodies also reflect separate periods of gold metallogenesis. In view of all single grain ages obtained, the three groups of ages 248–216, 213–200 and 183–157 Ma indicate that episodes of gold metallogenesis in the Jiapiguo belt

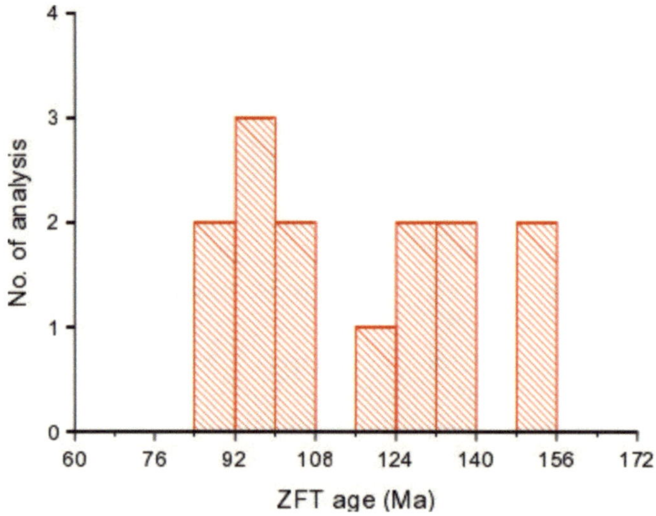

Figure 3. Histogram of zircon fission track (ZFT) ages obtained from ores from different deposits in the Jiapiguo gold belt. The three age groups of 86–106, 120–138 and 153–155 Ma indicate Indosinian and Yanshan epochs.

occurred during the Indosinian to Yanshanian epochs. However, some samples yielded old ages of 2342–2674 Ma, reflecting ore (metal) provenance from Precambrian rocks.

The obtained zircon fission track ages ranging from 86 ± 5 to 155 ± 11 Ma all fall into the Yanshanian epoch. According to distribution of the age data, these ages could be divided into three groups: 155–153, 138–120 and 106–88 Ma. Because zircon fission track age has a lower retention temperature than a corresponding zircon U–Pb age, the zircon fission track age group of 155–153 Ma likely corresponds to the zircon U–Pb age group of 170–166 Ma. The other two zircon fission track age groups-138-120 and 106–88 Ma likely represent younger periods of gold metallogenesis in the Jiapiguo belt.

Biotite from an altered host rock yielded a well-defined plateau $^{40}Ar/^{39}Ar$ age of 228.9 ± 2.2 Ma. K-feldspar from altered host rocks yielded total gas $^{40}Ar/^{39}Ar$ ages of 217.5 ± 22.8 and 197.9 ± 25.1 Ma, and another plateau $^{40}Ar/^{39}Ar$ age of 170.3 ± 2.6 Ma. These age data coincide well with the two older zircon U–Pb age groups mentioned above. Because altered minerals (biotite and K-feldspar) were formed during gold mineralization, their $^{40}Ar/^{39}Ar$ ages likely correspond to certain periods of gold metallogenesis in the Jiapiguo belt.

Therefore, the present age data obtained using three different methods reveal five periods of gold metallogenesis 230.0–228.9, 208, 170–153, 138–120 and 106–88 Ma in the Jiapiguo belt. The three middle groups of ages are the most likely plausible in view of the tectono-magmatic history of the North China Craton where the Jiapiguo and other gold belts in eastern China are situated. The multiple periods of gold metallogenesis in the Jiapigou belt were likely responsible for the formation of large gold deposits in the region, besides role of Precambrian rocks as plausible metal sources.

As a remarkable outcome of the present study, the multiple episodes of gold metallogenesis defined from the age data strongly correspond with ages of certain tectonic-magmatic events that affected the northern margin of the North China Craton as reported from previous studies. Therefore, the age data obtained from zircon U–Pb dating, zircon fission track dating and $^{40}Ar/^{39}Ar$ dating in the present study reflect not only periods of gold metallogenesis but also periods of corresponding tectonic-magmatic activities that resulted in the super-large gold deposits in the Jiapiguo belt.

During the Indosinian epoch, gold mineralization in the Jiapugo belt can be associated with bidirectional continental subductions of the North China Craton due to its collision with the Angara Craton in the north and the Yangtze Craton in the south that resulted in continental oversteps and an evolutionary series of granitoids. The collision of the Siberia and Mongolia blocks with the North China Craton during Late Triassic-Early Jurassic (about 180–150 Ma) led to closure of the Mongolia-Okhotsk Ocean, inducing Cretaceous metallogenic processes nearly at the same time. The Late Yanshanian tectonic-magmatic-mineralizing events in the Jiapigou gold belt were mainly controlled by the subduction of the Pacific plate. The metallogenic episodes were contemporaneous with changes in a drift direction of the subducting Pacific plate. The subduction-related magmatisms, which were controlled by deep-seated regional faults (e.g., the NE-NNE-trending Tanlu Fault), mainly originated in the middle crust and/or near the Moho discontinuity from which ore-forming fluids were likely extracted [29].

5. Preservation of ore deposit

A useful approach is to use low-temperature thermochronology methods to constrain the unroofing/exhumation history of a region. In regions unaffected by extension, the primary process driving rock exhumation will be erosion. Take Harizha-Halongxiuma Cu-Mo polymetallic ore deposit for example, it is located in the eastern Central East Kunlun zone (**Figure 4b**). The East Kunlun Range is a complex lithospheric left-lateral strike-slip fault system that located in the northeastern Tibetan Plateau separating the Bayan Har-Songpan Ganzi to the south and the Qaidam basin to the north. Cu-Mo orebodies occur mainly with the granite porphyry, granitoid diorite and monzogranite. Some Cu ore bodies also occur in the contact zones and fracture zones.

5.1. Sample preparation and analytical technique

The collected rock samples were subjected to crushing, grinding and sieving. Both apatite and zircon concentrates were separated from the crushed rock samples using conventional magnetic and heavy-liquid separation techniques.

5.1.1. Age results

A total of six apatite and six zircon fission-track results were successfully completed for all the samples. In addition, two apatite (U-Th)/He ages were obtained. The results are presented in **Tables 2** and **3**. Samples with chisquare >5% means all grains in it formed within the same tectonic/thermal event. The apatite fission-track ages narrowly range from (102 ± 4) Ma to (87 ± 6) Ma with the mean track lengths ranging from (11.9 ± 2.3) μm to (12.6 ± 1.6) μm and the zircon

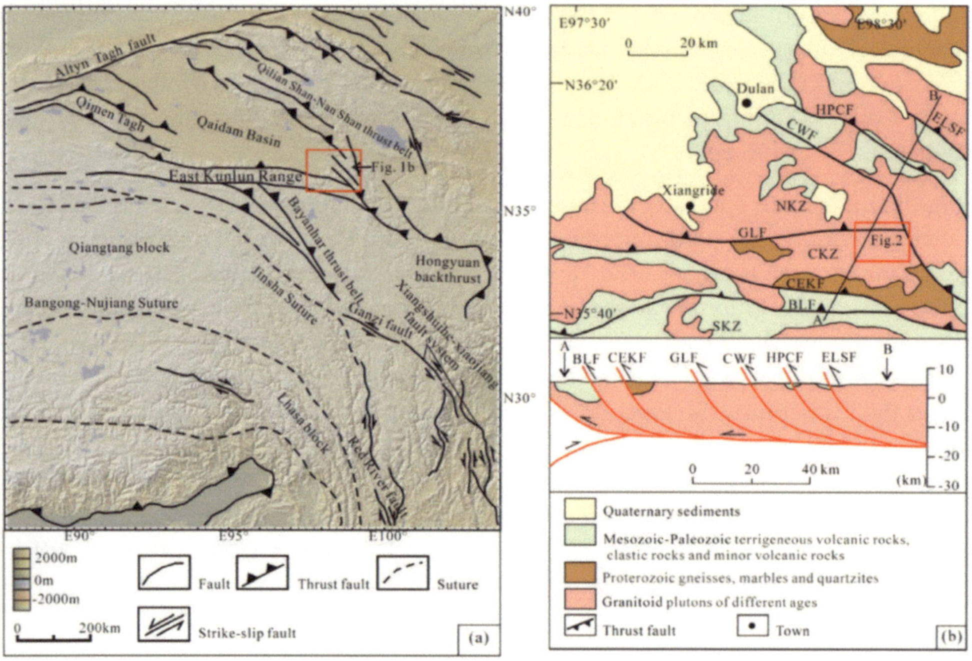

Figure 4. Map showing the location and geological setting of study area. (a) Map of the Tibetan plateau with main large-scale faults and terranes. (b) Sketch map of the eastern east Kunlun range, indicating the Elashan fault (ELSF), the Hacipu fault (HCPF), the Chahanwusu fault (CWF), the Goli fault (GLF), the central east Kunlun fault (CEKF) and the Boluoer fault (BLF). The section A-B is shown beneath the sketch map.

fission-track ages are ranging from (205 ± 14) Ma to (142 ± 7) Ma. The apatite (U-Th)/He dating test results of every grains provides a group of ages ranging from (68 ± 4.1) Ma to (52 ± 3.1) Ma with the mean ages of samples ranging from (60 ± 3.6) Ma to (56 ± 3.4) Ma.

5.1.2. Geothermal history

The HeFTy software [30] was used to model the thermal history based on the track length distribution and fission-track age data. Modeling of the fission-track data was based on our double dated samples with apatite and zircon fission-track dating. Time–temperature history was calculated using the annealing model of [31] and c-axis projected length [32]. The initial apatite fission-track length for this modeling was 16.3 μm. The modeling results are shown in **Figure 5**.

In **Figure 5**, all the temperature-time paths show a remarkable multi-stage pattern, including two rapid cooling stages and one relatively stable period in the Harizha-Halongxiuma ore deposit area since the Mesozoic. The time-temperature paths tend to be flat or have a low slope between approximately 100–60 Ma but become much more oblique before approximately 100 Ma and after approximately 60 Ma. The mean apatite (U-Th)/He ages are distributed around 60–56 Ma, which means that a remarkable decrease in temperature took place during the Paleocene. This is perfectly consistent with the thermal history in **Figure 5**. Besides

Sample/ mineral	Elevation	Rock type	n	No. of field of views	Spontaneous $\varrho_s(N_s)$	Induced $\varrho_i(N_i)$	Dosimeter $\varrho_d(N_d)$	P (χ^2) (%)	Central age (Ma) (± 1σ)	Mean track length (± 1σ)
XN12 + 15-Ap	4147	Crystal tuff	30	108	3.836(893)	8.583(1998)	11,682(8116)	98.7	102 ± 4	12.4 ± 1.9
XN12 + 15-Zr	4147	Crystal tuff	15		100.866(2102)	53.12(1107)	19.437(10030)	84.8	166 ± 9	
XN23-1-Ap	4358	Granite	30	101	3.641(958)	8.771(2308)	11.462(8116)	83.4	93 ± 6	12.4 ± 1.9
XN23-1-Zr	4358	Granite	27		112.321(3754)	75.908(2537)	21.385(10030)	83.9	142 ± 7	
XN26-Ap	4021	Quartz vein	30	101	3.973(1070)	10.563(2273)	11.02(8116)	68.0	101 ± 6	12.6 ± 1.6
XN26-Zr	4021	Quartz vein	8		181.112(1086)	91.556(549)	23.218(10030)	98.4	205 ± 14	
XN28-Ap	4094	Quartz vein	34	103	4.178(881)	10.126(2135)	10.8(8116)	96.5	87 ± 6	12.2 ± 1.7
XN28-Zr	4094	Quartz vein	7		169.125(1330)	88.377(695)	22.875(10030)	94.9	196 ± 12	
XN30-2-Ap	4138	Granite	28	102	4.074(1060)	8.555(2226)	10.579(8116)	83.8	98 ± 6	12.5 ± 1.9
XN30-2-Zr	4138	Granite	28		197.025(7166)	96.505(3510)	22.187(10030)	73.2	203 ± 9	
XN33-1-Ap	4069	Diorite	28	104	5.482(443)	11.782(952)	11.02(8116)	75.5	100 ± 8	11.9±2.3
XN33-1-Zr	4069	Diorite	28		181.299(6398)	93.257(3291)	21.843(10030)	44.4	190 ± 9	

Note: Track densities (ϱ) are as measured and are (10^5 per cm^{-2}); number of grains counted (n) are shown in parentheses. P(χ^2) is chi-square probability. Apatite ages calculated with ζ (zeta) = 392 ± 18.7 (yr cm^2/tr) and zircon ages calculated with ζ (zeta) = 90.9 ± 3.5 (yr cm^2/tr).

Table 2. Fission-track ages of apatite from the Harizha-Halongxiuma mine area using an external detector method and zeta calibration approach.

Sample	He (nmol/g)	U (ppm)	Th (ppm)	Sm (ppm)	FT	He age (Ma)	Error
DL14-3_1	0.1452	5.1	17.6	105	0.66	68	4.1
DL14-3_2	0.1938	8.9	26.2	144	0.66	59	3.5
DL14-3_3	0.2424	11.3	42.0	177	0.65	52	3.1
Mean	0.1938	8.4	28.6	142.0	0.66	60	3.6
DL25 + 26_1	0.7024	11.7	38.3	171	0.74	56	3.4
DL25 + 26_2	0.5432	6.9	25.8	110.0	0.76	55	3.3
Mean	0.6228	18.6	32.1	141	0.75	56	3.4

Note: The estimated analytical uncertainty for He ages is about 6% (2σ).

Table 3. Apatite (U-Th)/He ages of Harizha-Halongxiuma mine area performed by furnace heating for He extraction and ICP-MS for U-Th determinations at the London Geochronology Centre.

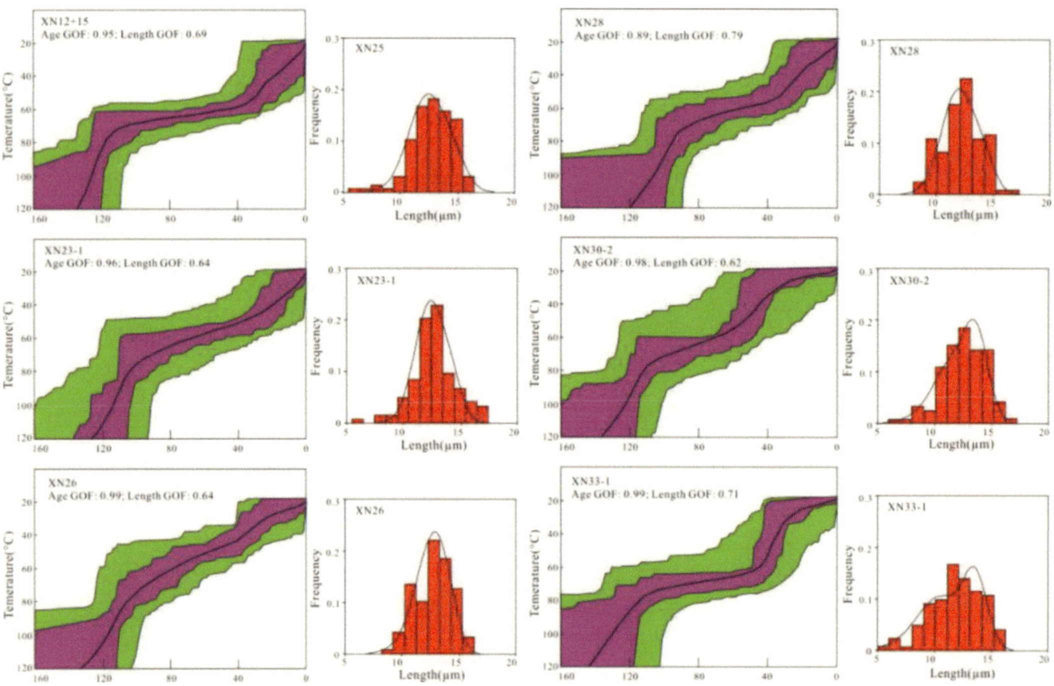

Figure 5. The inverse modeling results of samples in Harizha-Halongxiuma mine area. The time-temperature paths are modeled by the apatite fission-track data and distributions of apatite fission-track lengths with HeFty program by Ketcham [30]. The present day temperature is set as a constant value of 20°C. The dark gray area corresponds to the range of good fits, and the light gray area corresponds to the range of acceptable fits. The bold line in the dark gray area represents for the best modeled path. The Kolmogorov-Smirnov test was used to determine the goodness-of-fit of the solutions to the real data.

the temperature-time paths in **Figure 5**, the geothermal history with the temperature above 110°C can also be deduced. Thermal history with temperatures ranging between 210 and 110°C can be constrained by zircon and apatite fission-track ages because the temperature of samples can be defined as zircon fission-track closure temperature when samples pass through the top of part annealing zone. Thus, samples in this chapter had undergone a rapid cooling stage after approximately 200 Ma.

Taking all the temperature–time paths and the apatite (U-Th)/He ages into consideration, we conclude the geothermal history of the Harizha-Halongxiuma ore deposit area as a three-stage pattern as follows: (1) the first rapid cooling stage during approximately 200–100 Ma; (2) a subsequent stable stage during approximately 100–60 Ma; and (3) the second rapid cooling stage after 60 Ma.

5.2. Discussion

5.2.1. Exhumational processes and mechanism

Thermal history should be consistent with the tectonic setting and geodynamics. The thermal history of Harizha-Halongxiuma ore deposit area should be mainly affected by a series of collisions and matching between different blocks of Tibet Plateau.

Our ZFT and AFT data obtained in Harizha-Halongxiuma mining area indicate a decrease of approximately 100°C (the closure temperature difference between the ZFT and AFT) in temperature between about 205–142 and 102–87 Ma with the amount of uplift-induced exhumation of 3.60 km. This large-scale uplift and uplift-induced exhumation during the Early-Middle Mesozoic was consistent with the transpressional thrust nappeing system in the East Kunlun Range.

As shown in **Figure 5**, the rapid cooling stage ended during about 120–100 Ma followed by the onset of a stable stage with relatively low cooling rate. The Neo-Tethys Ocean closed in about 60–40 Ma and the India plate started collide with Eurasian plate. Even though the collision initially took place on the southern edge of Tibet, the effect of the collision and subduction propagated northward and led to the large-scale uplift in the whole Tibetan Plateau. Correspondingly, both of the (U-Th)/He data and inversing modeling temperature-time paths in this chapter indicate that there is a rapid cooling event took place during 60–40 Ma and resulted in a large amount of uplift-induced exhumation in Cenozoic.

5.2.2. Preservation of the ore deposit

The Harizha ore deposit area underwent Mesozoic rapid cooling event and recorded ZFT ages much later than the Halongxiuma ore deposit area. As a result, the uplift-induced exhumation of each deposit area is also different since the Mesozoic. The ore bodies in Harizha-Halongxiuma ore deposit area have experienced a three-stage (230–100, 100–60 and 60–0 Ma) cooling path since the mineralization epoch with the cooling rates of 1.26, 0.50 and 1.50°C/Myr, respectively for each stage. Based on the geothermal gradient of 35°C/km, the amount of uplift-induced exhumation of each stage is 4.68, 0.57 and 2.57 km, respectively. The total amount of uplift-induced exhumation in study area is 7.82 km. Microthermometric studies of fluid inclusions in ore rock samples from study area revealed that the burial depth of ore deposits mineralization in the Harizha-Halongxiuma ore deposit area ranges from 6.5 to 8.6 km [33]. The fact that 7.82 km of exhumation is larger than the minimum burial depth of the ore deposit in Harizha-Halongxiuma ore deposit area indicates that the upper part of the ore bodies had already been partially denuded. Fortunately, the maximum burial depth of mineralization of

8.6 km is larger than the thickness of exhumation, which likely means that part of ore bodies are still buried under the Earth's surface. Generally, the exhumation thickness since the main mineralization epoch indicates that the ore bodies in Harizha-Halongxiuma ore deposit area have been nearly half eroded. Ore bodies of approximately 0.8 km in depth are still well preserved under the surface.

In Harizha-Halongxiuma mine area, the copper and molybdenum deposit belts are always observed as developing with/within the fracture zones. Y1-Y2 section was designed to discover the relationship between ore bodies and the host strata. Attention had been paid to define whether they had undergone the same denudation history. Based on our dating results and the modeling of temperature-time paths, sample XN12 + 15 started to experience Cenozoic rapid cooling in approximately 40 Ma but the apatite (U-Th)/He ages of DL14-3 shows a geothermal story of rapid cooling in about 68–52 Ma. Thus, the cooling onset of DL14-3 was clearly early than that of XN12 + 15. We infer that a thrusting movement happened in the Early Paleogene had brought DL14-3 from a much deeper layer, and the time was recorded by the helium due to high-rate exhumation. The ore bodies hosted in the Triassic strata were partially eroded along with the hanging wall (**Figure 6a**). For section X1-X2, four samples, such as XN26, XN28, XN30-2 and XN33-1, had undergone nearly the same exhumation history as their zircon and apatite fission-track ages fluctuated within a relatively narrow range. This indicates that the Permian–Triassic intrusions where the four samples were collected have undergone uniformed geothermal history and were exhumed to the Earth's surface together. In addition, the southwestward thrusting made the hangingwall underwent much thicker exhumation comparing with footwalls to the southwest (**Figure 6b**). The molybdenum mineralization was suggested to mainly form with/within the granitoid intrusions near the location of our samples. It is likely the rapid exhumation in Cenozoic that eroded the cover of the Mo-orebodies and made the orebodies outcropped on the Earth's surface.

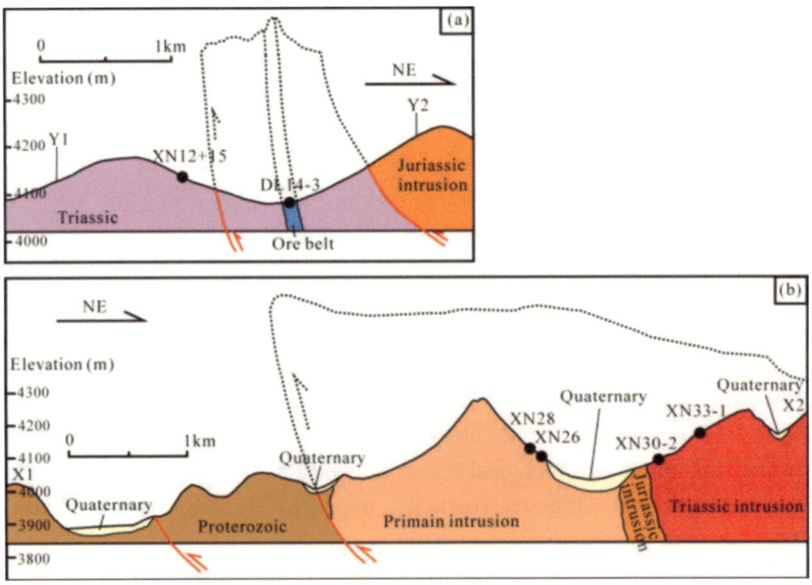

Figure 6. Cutaway views of the section Y1-Y2 (a) and section X1-X2 (b) in Harizha-Halongxiuma mine area. The ideally inferred exhumed sections are shown in dash area.

http://dx.doi.org/10.5772/intechopen.72277

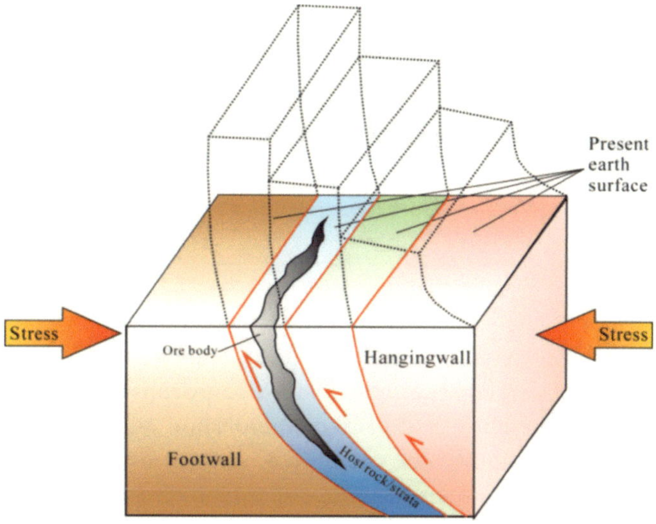

Figure 7. The ideal exhumation model for the ore bodies in compressional orogenic belt. Dash parts represent eroded stratum/intrusions/orebodies.

The analysis of exhumation history in Harizha-Halongxiuma indicates that the thrust faults in mine area exert profound influence on the preservation of ore bodies. The hangingwalls were extruded out by compression and were uplifted higher than the regional average elevation. The bulge of the Earth's surface would undergo a strong weathering process and would be eroded and exhumed rapidly. The ore bodies hosted in the hangingwall would also be eroded much more easily than that in the footwall. Thus, an ideal model is presented to outline the preservation and exhumation history of ore bodies in compressional orogenic belts (**Figure 7**). As shown in **Figure 7**, the collisions and matching between continental blocks exert compressional stress in mine areas and resulted in the development of thrust faults. The convergency between the hangingwalls and the footwalls would lead to the continuous extrusion of hangingwalls, which would result in the uplift of hangingwalls. The subsequent weathering on the bulge of the Earth's surface would result in the high-rate uplift induced exhumation. As a result, the ore bodies formed within the host rocks/stratum would be eroded due to the subsequent strong planation. The ore bodies hosted in the footwalls would be well preserved than that hosted in the hangingwalls.

So, it could be said that Harizha-Halongxiuma ore deposit area have several characteristics:

1. Apatite fission-track ages of the samples collected from the Harizha-Halongxiuma ore deposit area were obtained as (102 ± 4)-(87 ± 6). The zircon fission-track ages are ranging from (205 ± 14) to (142 ± 7) Ma. The apatite (U-Th)/He ages range from 59 to 54 Ma.

2. Six confidently modeled thermal history diagrams were obtained, and they show similar characteristics, such as rapid cooling in approximately 200–100 and 60–0 Ma and a stable period of thermal history from 100 to 60 Ma.

3. The respective average cooling rates and inferred uplift-induced exhumation for the three stages were calculated as 1.26°C/Myr and 3.60 km for the first rapid cooling stage, 0.50°C/Myr and 0.57 km for the subsequent slow cooling stage, and 1.50°C/Myr and 2.57 km for the second rapid cooling stage.

4. The exhumation of the Harizha-Halongxiuma ore deposit area since about 230 Ma was ideally calculated as 7.82 km, and the ore deposits in the study area are considered to be partially preserved.
5. An ideal exhumation model for ore bodies in compressional orogenic belt was established [34].

6. Method summary

Strong tectonic uplifting cause the gold polymetallic metallogenic system has been subjected to intense change and reform. For example, alkali-rich porphyry type of gold ore deposits in Western Yunnan Province have multi-stage of evolution and obvious diachronism, and go through conservation-changes of the ore bodies, such as uplifting-denudation, burying, shearing, missing, mineralizing superposition, supergene enrichment, long-distance migration, leaching, depletion, etc.

Same as mineralization, the conservation and changes of ore deposits would be the important part of mineral deposit geology. Research on the conservation and changes can deeply reveal the mineralization, evolution history, preservation condition, change factor, distribution regularity and hidden deposit, which are favorable for prospecting and xploration work.

6.1. Uplift and exhumation

The uplift and exhumation of the geologic bodies is the key factor of the affection in the preservation and the varition of the deposit. Uplift and denudation can directly lead to the formation of ore deposits, such as the Gandese batholithic zone in Tibet, with about 15Ma after uplift, along with the extension and collapsion of the earth crust, ore magma emplace and form the porphyry copper deposits [35]. More importantly, uplift and denudation is the key factors that the ore deposits can preserve after the formation of ore deposits [36, 37]. Mesozoic Tuwu copper deposits have been preserved in Xinjiang, for example, since differential upliftings occurred in Tuwu copper district and surrounding areas (north of Kanggur fracture), Tianshan Mountains, and then the denudation area changed into sedimentary area [38]. Bei Ya large gold deposits in Yunnan and Huayuan large lead-zinc mine in Hunan, have the low denudation degree and good preservation characteristics.

6.2. Research techniques and content

6.2.1. Research techniques

The principle of using fission track method to determine the geological age and radioactive isotope chronology are both on the basis of the percentage and the rate of disintegration in samples to determine the age of the geological samples.

The difference between them is that the fission track measurement is the radiation damage effect of fission, according to the number of 238 U in mineral, which comes from the spontaneous fission and fission time which is from spontaneous fission rate, that is the fission track age, while the other method is measuring the product of the decay, that is according to the content of parent and daughter isotopes and the maternal isotopes decay rate, and determine the length of the decay time.

Different minerals have different fission closure temperature. The commonly used minerals in fission track are apatite, zircon and SPAR. The apatite is about 110°C [39, 40], and the zircon is 250°C [41, 42] and the SPAR is 340°C [43, 44]. The thermal evolution of different minerals in different minerals in the same mine belt can obtain the thermal evolution process of higher temperature. Hydrothermal deposit metal mineralization mainly occurs in the high temperature (350–250°C), which in the formation of quartz-sericitization, and low temperature (250–150°C) that in the formation stage of altered minerals such as kaolinization and montmorillonite, propylitization, so the fission track age of the zircon and sphene can represent metallogenic epoch. It is certainly better to combine the methods of the ^{40}Ar-^{39}Ar thermochronology, the zircon U–Pb chronology, (U-Th)/He thermochronology, and the age of the res-os.

Anyhow, basen on geological characteristics and application of a variety of thermochronology methods, we quantitatively study the uplifting-denudation and exhumation histories of the each ore districts and ore belts in Jiage, inteprete the preservation and change conditions of different ore deposits, and provide evidence for regional metallogenic potential evaluation and deep prospecting prediction.

6.2.2. Research content

1. The basic geological study
2. The reasonable sampling system
3. Thermochronology test analysis
4. Quantitative study of uplift and exhumation
5. To establish the environment model and the prospecting forecast for ore deposits

6.3. Technical solution

1. Research area selection
2. Basic geological research
3. Sample collection
4. Single mineral separation
5. Thermochronology tests
6. Geological history simulation
7. Cooling rate

Another way is by using the formula, the cooling rate Cr:

$$Cr(^{\circ}C/Ma{-}1) = (Tm - Tsurf)/tm \quad (3)$$

Tm is the closure temperature of the mineral fission track, Tsurf is the temperature of the Earth's surface and tm is the age of the surface sample.

8. The uplift rate and uplift amplitude.

$$U = D + \triangle H + \triangle s.l. \quad (4)$$

The D is the denudation, and △H is the change in height, and △s.l. is the magnitude of sea level change.

The rise of the rock can also be expressed as:

$$U = \text{Surface Uplift} + \text{Erosion} \quad (5)$$

The "Surface Uplift" can be described as △E, or "Exhumation". This way, the Earth's surface increases by the same amount as u-△E. The relationship between the rise of the rock and the rise of the Earth's surface can be learned.

9. The rate of erosion.

Denudation is the degree of erosion of surface rock caused by erosion or tectonic action. Erosion causes sustained cooling, so there is an erosion rate:

$$\text{Denudation rate} = \text{cooling rate/geothermal gradient} \quad (6)$$

Second, the temperature gradient (Q/k) is calculated based on regional heat flux (Q) and heat conduction coefficient (k). The third is to collect data that have been obtained from previous generations. It is important to note that the geothermal gradient is a regional concept. Using the sample's age and elevation drawing diagram, if the correlation is better, it can reflect the denudation rate; therefore, it can be calculated at different stages of denudation rate.

For low-temperature thermochronology, especially (U-Th)/He method, topography and wavelength, hypsography, terrain elevation difference on the surface of isothermal surface are the influencing factors; among them, the main influence is wavelength and terrain. For this reason, the denudation rate of the request can be adjusted according to the 2D model of temperature stability distribution.

10. Denudation amount

The average denudation (△E) can be calculated based on the following formula:

$$\triangle E = (110 \pm 10^{\circ} C - Ts)/G + d \quad (7)$$

Ts is the ancient land surface temperature, G represents the paleogeothermal gradient, d is for fission track annealing with the bottom elevation and the surface elevation difference (d = fission track annealing with elevation – now at the bottom of the surface elevation).

The denudation can be calculated based on the rate of erosion and duration. In a certain period of time, its formula is as follows:

$$D = (TB - TO)/G \tag{8}$$

where D is the thickness of strata denudation (km), TB simulate the ancient temperature (°C), which corresponds the starting point for rapid cooling event, the temperature of the heat is the historical turning point, TO represent the fast cooling events at the end of the ancient temperature (°C) and G is on behalf of the geothermal gradient (°C/km).

11. Comprehensive analysis of uplift and exhumation.

Combining various quantitative data and qualitative data with the actual geological characteristics of the area, it is found that the period of metallogenic stage, the period of tectonic activity and the relation between the two are closely related; systematically, it summarizes the cooling rate of different mining areas, different parts, different time periods, the rate of growth, the rate of increase, the rate of denudation, denudation and total denudation. The evolution pattern of cooling, rising and denudation of the zone is revealed, and the evolution pattern of metallogenic and metallogenic is proposed.

12. Deposit conservation and change.

The formation depth of the deposit in the zone and the degree of denudation of the various ore deposits are obtained. The relationship between the depth and the denudation of the mineralization was summarized, and the whole process of the formation of different ore districts (beds) was probed. We identify various evolution regularities, establish the preservation-change models of the ore deposits and predict concealed ore deposits in different locations of different ore districts.

In a word, by studying a lot of hot chronology data is available, and explore the metallogenic period of time, tectonic activity and the relationship between the two, find out the mineralization and metallogenic after preservation changes during uplift denuding conditions, quantitative give the uplift rate of mining area, different sections and the main ore deposit, uplifting amplitude and stripped, reveal the regularity of mineralization and metallogenic after preservation changes during evolution and history, to set up new ore deposit geology—preserve the environment model, predict different mining area, the different locations of concealed ore deposits can be output depth, and give deposit which may have had been denudated to its extent [45].

Author details

Wanming Yuan* and Ke Wang

*Address all correspondence to: ywm010@yahoo.com

Institute of Earth Science, The China University of Geosciences, Beijing, China

References

[1] Wanming Y, Xuanxue M, Aikui Z, et al. Fission track thermochronology evidence for multiple periods of mineralization in the Wulonggou gold deposits, eastern Kunlun Mountains, Qinghai Province. Journal of Earth Science. 2013;**24**(4):471-478

[2] Silk ECH, Barnes RS. Examination of fission fragment tracks with an electron microscope. Philosophical Magazine. 1959;**4**:970-972

[3] Price PB, Walker RM. Chemical etching of charged-particle tracks in solids. Journal of Applied Physics. 1962;**33**:3407-3412

[4] Price PB, Walker RM. Observation of fossil particle tracks in natural mica. Nature. 1962; **196**:732-734

[5] Fleischer RL, Price PB. Glass dating by fission fragment tracks. Journal of Geophysical Research. 1964;**69**:331-339

[6] Fleischer RL, Price PB, Walker RM. Effects of temperature, pressure and ionization on the formation and stability of fission tracks in minerals and glasses. Journal of Geophysical Research. 1965;**70**:1497-1502

[7] Guo S, Wanming Y, Baoliu C, et al. Track fading and its applications in archaeology, tectonics and geothermal chronology in China. Journal of Earth Science. 2013;**24**(4):645-651

[8] Kang TS, Wang SC. The Fission Track Analysis Method in the Study of Geothermal Histories. Beijing: Science Press; 1991 (in Chinese)

[9] Yang T, Jianxiang Z, Wanming Y, et al. Proofs on feasibility of measuring geologic age by a thermal analysis method. International Journal of Environmental Science. 2016;**1**:97-105

[10] Jiang Y, Chang H. Apatite(U-Th)/He dating a review. Acta Petrologica et Mineralogica. 2012;**31**(5):757-766 (in Chinese)

[11] Chang Y, Xu CH, Zhou ZY. (U-Th)/He dating method: α-ejection influence and correction. Advances in Earth Science. 2010;**25**(4):418-427 (in Chinese)

[12] Xu CH, Zhou ZY, Chang Y. Genesis of Daba arcuate structural belt related to adjacent basement upheavals: Constraints from fission-track and (U-Th)/He thermochronology. Science China Earth Sciences. 2010;**40**(12):1684-1696 (in Chinese)

[13] Yang TS. Feasibility of probing solid state nuclear tracks by thermal analysis method. Chinese Science Bulletin. 2007;**52**(4):380-383 (in Chinese)

[14] Yang TS, He SR, Li TX, et al. Measurement of solid state nuclear tracks in apatite by thermal analysis method. Chinese Science Bulletin. 2009;**54**(17):2495-2499 (in Chinese)

[15] Yang TS, Shi YQ, Li YG, Wang DW, et al. Research program of determination of geological age by thermal analysis method. Journal of Earth Science. 2013;**24**(4):657-662

[16] Bellemans F, De Corte F, Van Den Haute P. Composition of SRM and CN U-doped glasses: Significance for their use as thermal neutronfluence monitors in fission track dating. Radiation Measurements. 1994;**24**(2):153-160

[17] Hurford AJ, Green PF. A users'guide to fission-track dating calibration. Earth and Planetary Science Letters. 1982;**59**:343-354

[18] Hurford AJ. Standardization of fission track dating calibration: Recommendation by the fission-track working group of the I.U.G.S. subcommission on geochronology. Chemical Geology. 1990;**80**:171-178

[19] Galbraith RF. On statistical models for fission track counts. Mathematical Geology. 1981; **13**:471-488

[20] Yuan W, Bao Z, Dong J, et al. Zircon and apatite fission track analyses on mineralization ages and tectonic activities of Tuwu-Yandong porphyry copper deposit in northern Xinjiang, China. Science in China Series D: Earth Sciences. 2007;**50**(12):1787-1795

[21] Green PF, Duddy IR, Laslett GM, et al. Thermal annealing of fission tracks in apatite, 4. Quantitative modeling techniques and extensions to geological timescales. Chemical Geology. 1989;**79**:155

[22] Wanming Y, Wang S, Shengrong L, et al. Apatite fission track dating evidence on the tectonization of Gangdese block, south Qinghai-Tibetan plateau. Chinese Science Bulletin. 2002;**47**(3):240-244

[23] Yuan WM, Zhang XT, Dong JQ, et al. A new vision of the intracontinental evolution of the eastern Kunlun Mountains, Northern Qinghai-Tibet plateau, China. Radiation Measurements. 2003;**36**(2003):357-362

[24] Wanming Y, Zheng Q, Bao Z, et al. Zircon fission track thermochronology constraints on mineralization epochs in Altai Mountains, northern Xinjiang, China. Radiation Measurements. 2009;**44**(2009):950-954

[25] Copeland P, Harrison TM, Yun P. Thermal evolution of the Gangdese batholith, southern Tibet: A history of episodic unroofing. Tectonics. 1995;**14**:223-236

[26] Harrison TM, Copeland P, Kidd WSF. Activation of the Nyainqentanghla shear zone: Implications for uplift of the southern Tibet plateau. Tectonics. 1995;**14**:658-676

[27] Yuan Wanming, Hou Zhengqian, Li Shengrong, et al. Fission track evidence on thermal history of Jiama polymetallic ore district, Tibet. Science in China (Series D), 2001; **44**(Sup):139-145

[28] Wanming Y, Bao Z, Dong J, et al. Zircon and apatite fission track analyses on mineralization ages and tectonic activities of Tuwu-Yandong porphyry copper deposit in northern Xinjiang, China. Science in China (Series D). 2007;**50**(12):1787-1795

[29] Deng Jun, Yuan Wanming, Emmanuel John Muico Carranza, et al. Geochronology and thermochronometry of the Jiapigou gold belt, northeastern China: New evidence for multiple episodes of mineralization. Journal of Asian Earth Sciences. 2014;**89**:10-27

[30] Ketchma RA. Forward and inverse modeling of low-temperature thermochronometry data. Review in Mineralogy and Geochemistry. 2005;**58**(1):275-314

[31] Ketchman RA, Carter A, Donelick RA, Barbarand J, Hurford AJ. Improved measurement of fission-track annealing in apatite using c-axis projection. American Mineralogist. 2007; **92**:789-798

[32] Ketchman RA, Carter A, Donelick RA, Barbarand J, Hurford AJ. Improved measurement of fission-track annealing in apatite. American Mineralogist. 2007;**92**:799-810

[33] Xu QL. Study on Metallogenesis of Porphyry Deposit in Eastern Kunlun Orogenic Belt, Qinghai Province. Changchun, China: Department of Earth Sciences, Jilin University; 2014 (in Chinese)

[34] Feng YL, Yuan WM, Tian YT, et al. Preservation and exhumation history of the Harizha-Halongxiuma mining area in the East Kunlun Range, Northeastern Tibetan Plateau, China. Ore Geology Reviews 2017;**84**:116-133

[35] Qu XM, Hou ZQ, Khin Z, Mo XX, Xu WY, Xin HB. A large-scale copper ore-forming event accompanying rapid uplift of the southern Tibetan plateau: Evidence from zircon SHRIMP U–Pb dating and LA ICP-MS analysis. Ore Geology Reviews. 2009;**36**(1-3):52-64

[36] Márton I, Moritz R, Spikings R. Application of low-temperature thermochronology to hydrothermal ore deposits: Formation, preservation and exhumation of epithermal gold systems from the eastern Rhodopes, Bulgaria. Tectonophysics. 2010;**483**(3-4):240-254

[37] Schardt C, Large RR. New insights into the genesis of volcanic-hosted massive sulfide deposits on the seafloor from numerical modeling studies. Ore Geology Reviews. 2009;**35**(3-4):333-351

[38] Wu ZN, Huang JH, Yu SP, Muhtar ZR, Yang XR, Dili XT, Han WQ. Paleogeographic environment for reserving and forming of Tuwu deposit in eastern Tianshan Mountains, Xinjiang. Arid Land Geography. 2007;**30**(2):189-195 (in Chinese)

[39] Gleadow AJW, Duddy IR. A natural long-term track annealing experiment for apatite. Nuclear Tracks. 1981;**5**:169-174

[40] Kemp PJJ. Fission track analysis reveals character of collisional tectonics. Tectonics. 1989;**8**(2):169-195

[41] Yamada R, Tagami T, Nishimura S, Ito H. Anealing kinetics of fission tracks in zircon: An experimental study. Chemical Geology (Isotope Geoscience Section). 1995;**122**:249-258

[42] Wagner GA, Haute PVD. Fission track dating. Ferdinand Enke Verlag, Stuttgart. 1992:95-119

[43] Nagpaul KK. Fission track geochronology of India. In: Goswami JN, editor. Nuclear Tracks. Bangalore: India Academy of Sciences; 1982. pp. 53-65

[44] Gleadow AJW, Lovering JF. Geometry factor for external detectors in fission track dating. Nuclear Track Detection. 1977;**1**:99-106

[45] Yuan WM. Thermochronological method of revealing conservation and changes of mineral deposits. Acta Petrologica Sinica. 2016;**32**(8):2571-2578

Chapter 8

Statistical Approach to Mineral Engineering and Optimization

Mehmet Deniz Turan

Additional information is available at the end of the chapter

http://dx.doi.org/10.5772/intechopen.71607

Abstract

Mineral depositions are basic sources for obtaining metal production. Increasing metal demand based on increasing world population and decreasing grade value of mineral deposition make the evaluation to mineral processing more important, so that all metal production stages must be economical. Because of this important requirement, many researchers and practitioners have focused to the optimization of all processes. The optimization of metal production processes provide some advantages such as reducing the influence of experimental errors, statistical analysis, determining important parameters and trivial parameters, and measuring interactions between parameters. Although there are many design methods, choosing the most appropriate method is of great importance in terms of the results to be achieved. In this chapter, presumed experimental data about hydrometallurgical copper extraction accompanied by three parameters were applied to two different design models to compare the results.

Keywords: copper, mineral processing, hydrometallurgy, optimization, experimental design, statistically, leaching

1. Introduction

Different types of ore reservation are consumed away quickly depending on the growing population. Metal production, which is effective in making tools used to meet basic vital needs (cars, tools, buildings, ships, etc.), must also increase along with the increasing population. The main source of all these metals is the ores, and all ores in earth are known as nonrenewable sources. For these reasons, all stages of metal production such as characterization, mining activities, mineral processing, pyrometallurgical/hydrometallurgical/electrometallurgical routes, and final production methods must be economical. In the world, more than 10 billion tons of mine are produced. Seventy five percent of this production is derived from energy raw materials, 10% is from metallic mines, and 15% is from industrial raw materials. On the other hand, some metals have been taken to the status of critical metals by the European Commission (**Figure 1**).

As shown in **Figure 1**, it is clear that some metals have high supply risk status such as beryllium, cobalt, magnesium, germanium etc. These data show that the economy of metal production is very important and these conditions are pushed to researchers and producers for the recovery of secondary sources [2–5].

- At that rate, how a metal producing process is economical?
- How we can benefit more effectively from existing ore grade and reserves?
- Will ore reserves meet future needs?

The answer to all these questions can be given as optimization.

The optimization, which may be numeric or nonnumeric (known as categorical), must be carried out at every stage of the metal production process. Processes that need to be optimized from the beginning start with the ore detection, mineralization-mineralogy-petrography, geographical distributions and planning. The extraction of ore mining at optimum conditions, mining operations, transportation, marketing, and planning. Up on this stage, many steps such as truck tours, excavator cycling, casting distance (in open mines), reaching to ore with appropriate road, number of people and machines to work, and ore extraction period (underground mines) can be optimized for priority. The next stage is mineral processing, which contains ore crushing, grinding, sieving, handling, conditioning, enrichment (one or more of the appropriate methods such as magnetic-electrostatic enrichment, flotation, gravimetric methods, etc.), and dewatering. It is stated that 50–75% of total energy is consumed in crushing and grinding unit in a mineral processing plant [6]. This fact shows that the optimization of this

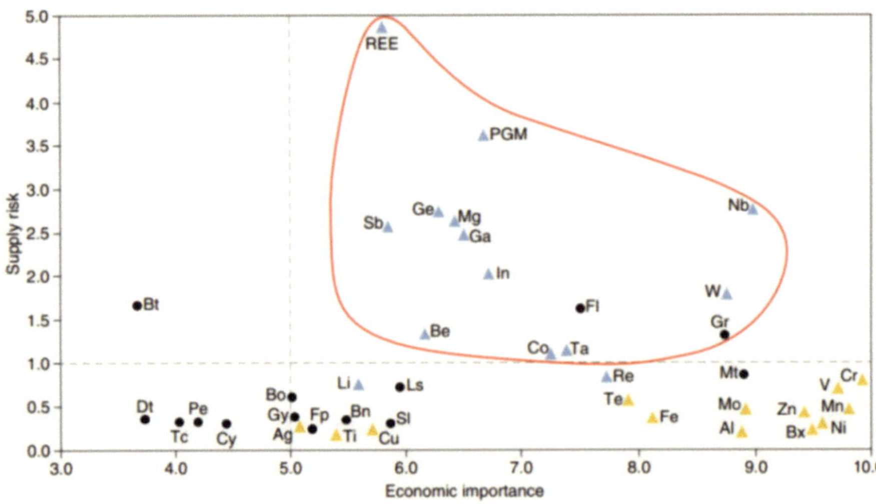

Figure 1. European Commission critical metal graphics [1].

stage (choosing of appropriate comminution and sieving machines, crushing-grinding time, treatment conditions, etc.) is quite important. The last stage is metal handling in which metallurgical methods are used, for instance, pyrometallurgy, hydrometallurgy, electrometallurgy, and casting. This stage also has a lot of substages so that some of them can be sorted as: temperature level, acid-base choosing, treatment time, solution volume, stirring conditions, electrode choosing, current applications, and casting conditions. We can roughly classify the phases as follows: ore detection, characterization, mining, mineral processing, metallurgy, and usage. How to be optimized? Which method should be used?

There are many packet computer software for this aim. Among them, one of the prominent is response surface methodology (RSM) that is used successfully in many engineering applications and especially in scientific articles. The use of response surface methodology is popular because of having easy-to-use interface and application advantages. The main advantages can be sorted as an opportunity of easy optimization, reaching maximum information by less experimental data, an opportunity of changing significant parameters simultaneously, determination of interaction between parameters, and easy elimination of insignificant parameters.

2. Experimental approach

In any process involving the metal production process, parameters that have an effect on metal recovery are primarily estimated. Then, it is aimed to determine the most favorable conditions by studying the parameters in a certain interval. However, while a single parameter is examined as a variable in traditional leaning studies, it is seen that other effective parameters are kept constant at a certain value. Furthermore, in such studies, it is only possible to define any model or model equation to represent the experimental method and only on special conditions. For example, in a hydrometallurgical study, it seems that the interaction of leaching parameters cannot be accounted, since the parameters investigated for activity on metal recovery are handled one by one. On the other hand, there is a need for a large number of experimental studies in a study in which all the parameters are examined.

2.1. RSM experimental design

Response surface methodology (RSM) was first defined and improved in 1951 by Box and Wilson. RSM is created with model regression analysis. Box and Wilson have setup experiments with the aim of reaching the point where the answer variable has the maximum value on the answer surface with the smallest possible number of observations. For this purpose, they have compared some experimental schemes and identified composite experiments [7]. Response surface methods are created with the help of model regression analysis, and it is successfully used in many disciplines. Two or more factors use the response surface pattern; for example, time, temperature, and the effects of both on the result can be investigated and optimum values are found. The results can be expressed in three-dimensional graphics or as a contour graphics. Using a very small number of experimental combinations, it is possible to estimate factors and their combinations that are not actually tested [8, 9]. In this study, a Stat-Ease software package (version 6.0.10 trial) is used for data analysis.

RSM is a collection of mathematical and statistical techniques that are useful for the modeling and analysis of problems in which a response of interest is influenced by several variables and the objective is to optimize this response. For example, suppose that an engineer wishes to find the levels of temperature (X_1) and pressure (X_2) that maximize the yield (y) of a process. The process yield is a function of the levels of temperature and pressure,

$$y = f(X_1, X_2) + \varepsilon \tag{1}$$

where ε represents the noise or error observed in the response y. If we denote the excepted response by $E(y) = f(X_1, X_2) = \eta$, then the surface is represented by

$$f(X_1, X_2) = \eta \tag{2}$$

In most RSM problems, the form of relationship between the response and the independent variables is unknown. Thus, the first step in RSM is to find a suitable approximation for the true functional relationship between y and the set of independent variables. If the response is well modeled by a linear function of the independent variables, then the approximating function is the first-order model,

$$y = \beta_0 + \beta_1 X_1 + \beta_2 X_2 + \ldots + \beta_k X_k + \varepsilon \tag{3}$$

If there is a curvature in the system, then a polynomial of higher degree must be used, such as the second-order model,

$$y = \beta_0 + \sum_{i=1}^{k} \beta_i X_i + \sum_{i=1}^{k} \beta_{ii} X_i^2 + \sum_{i<j}\sum \beta_{ij} X_i X_j + \varepsilon \tag{4}$$

Almost all RSM problems use one or both of these models [10].

Statistical design, also known as experimental design, is the methodology of how to plan and conduct experiments to obtain maximum information with a minimum of experiments. The following questions should be answered before any design selection is made to investigate variable parameters. All of these questions will have to be answered satisfactorily before the experiment is performed.

- What method of data analysis should be used? How will the impact of factors and response value be measured? How many factors will be effective on the response value? How many factors will be considered simultaneously? How many experiments will be needed again?

All principles based on statistical evidence have been proven by a reliable method, supported by trial and error. The use of experimental design methods provides a number of advantages in designing such studies, if there are a large number of variables that are effective on the result in a study.

Engineering processes are complex systems that are influenced by many factors. Experimental design ensures that these systems are expressed by functions and that important interactions between a large number of active variables are revealed. Response variants (metal recovery in leaching, flotation yield in flotation process, etc.) are observed as experimental outputs, while independent variables in the design of experiments are checked. Simultaneous exchange of variables allows arriving at a result in less time and with less experimental effort compared to the conventional experimental method in which a single variable is changed. On the other hand, the most important advantage of experimental design is the simultaneous variation of several factors and the independent evaluation of each factor. There are several common steps compared to most experimental designs. The first step is to determine the problem to be solved. Determining the factors affecting the process is the second step. Thirdly, it is the study of factors in different combinations in an experimental study. Finally, the best combination is to choose.

3. Design types

There are many types of designs for use to design experimental works. These design types have become easy to understand with the improvement of appropriate computer software. It is important to which one design must be chosen so that it is related to the fact that the researcher dominates the work to be done. Because there are a variety of design methods, their design criteria and application areas exhibit variety. Some of these design types are listed below for design expert software [11],

√ Central composite design, √ Box–Behnken design, √ Three-level factorial design, √ Hybrid design, √ One-factor design, √ Pentagonal design, √ Hexagonal design, √ D-Optimal design, and √ User-defined design, etc.

Each design type has its own characteristics, which are usually related to the position of design points and the number of design points. Short information on some design types is given below.

3.1. Central composite design

Each numeric factor is varied over five levels: plus and minus alpha (axial points), plus and minus 1 (factorial points), and the center point (**Figure 2**). The biggest advantage of the central composite design (CCD) is that it allows experimental design outside the design points where the cube points are located. In addition, the presence of these points gives the rotational workability, which is why it is highly preferred by researchers. CCD is divided into three subdivisions as central composite circumscribed (CCC), central composite inscribed (CCI), and central composite face-centered (CCF).

3.2. Box-Behnken design

Each numeric factor is varied over three levels. If categorical factors are added, the Box-Behnken design will be duplicated for every combination of the categorical factor levels (**Figure 3**).

3.3. Three-level factorial design

Each numeric factor is varied over three levels. If categorical factors are added, the three-level factorial design will be duplicated for every combination of the categorical factor levels (**Figure 4**).

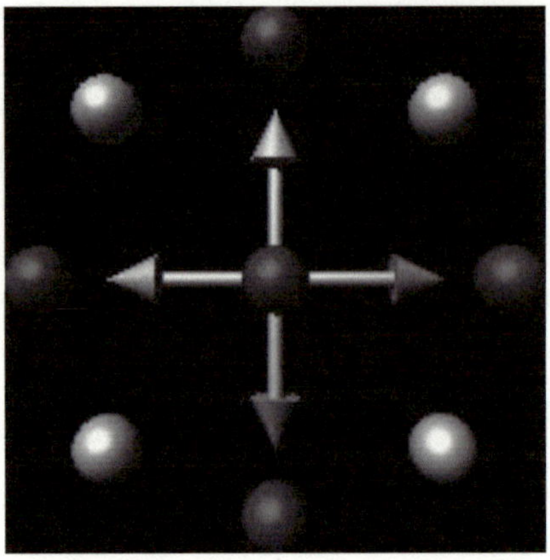

Figure 2. Central composite (CCD) design layout.

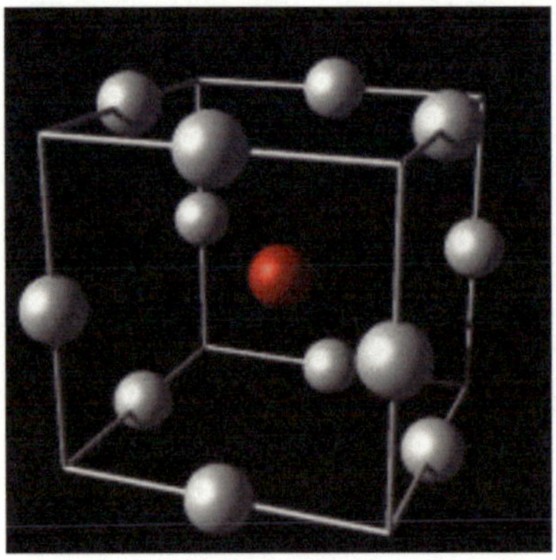

Figure 3. Box-Behnken design layout.

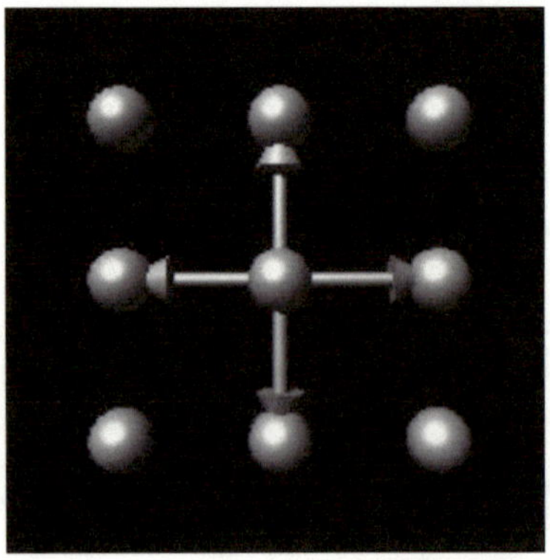

Figure 4. Three-level factorial design layout.

http://dx.doi.org/10.5772/intechopen.71607

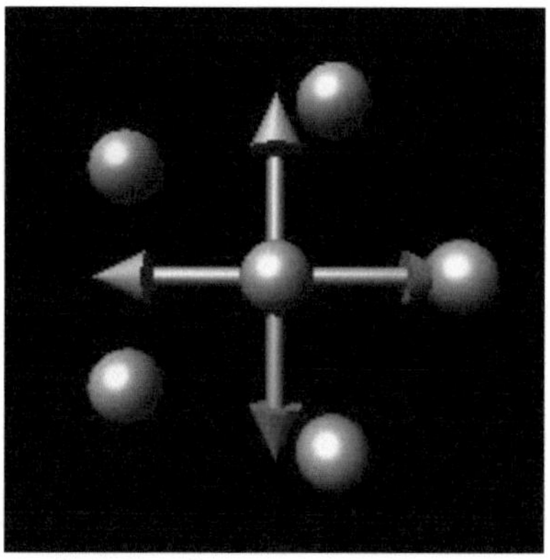

Figure 5. Pentagonal design layout.

3.4. Hybrid design

A minimal point design for three, four, six, or seven factors with five levels each. Because there is no replication, the lack of fit test is not available. These rotatable or nearly rotatable designs are better than a small central composite, but are still highly sensitive outliers or missing data.

3.5. One-factor design

One-factor design is a design for one numerical factor using three levels for a linear model, five levels for a quadratic model, seven levels for cubic models plus some replicated points.

The design will be duplicated for every combination of the categorical factor levels.

3.6. Pentagonal design

Design for two factors where factor A can have four levels and factor B has five levels. This minimal point design is extremely sensitive to outliers or missing data. The design will be duplicated for every combination of the categorical factor levels (**Figure 5**).

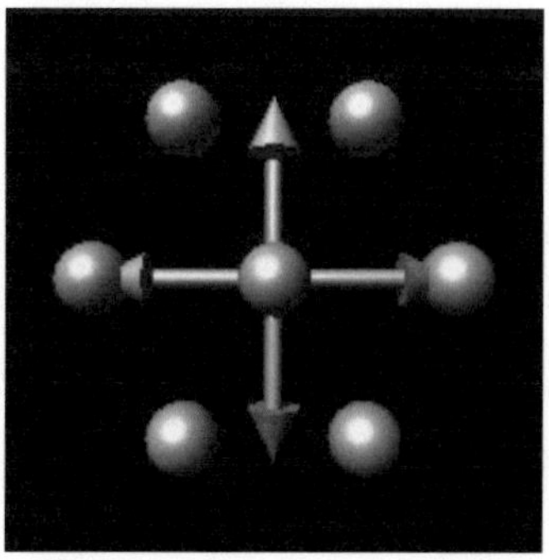

Figure 6. Hexagonal design layout.

3.7. Hexagonal design

Design for two factors where factor A has five levels and factor B has three levels. The design will be duplicated for every combination of the categorical factor levels (**Figure 6**).

4. Optimization of comparative sample hydrometallurgy study

Hydrometallurgy is an important part of the metal production process; it is becoming more important because it has many advantages such as being eco-friendly, easy operation, low energy depletion, and low cost. Recently, the planning of hydrometallurgy studies with RSM increases to examine the effect of many effective parameters [12–18]. The advantages of using response surface methods in such studies can be listed as follows:

- Reduce the influence of experimental errors.
- Allows statistical analysis.
- Helps determine important parameters and trivial parameters that need to be checked.
- Helps to determine and measure interactions between parameters.
- Allows for the best results to be searched within the examined intervals of the test parameters and to extrapolate the data.
- Enables you to draw graphs describing how variables relate to each other and to determine the values of variables that give optimum results.
- Allows the creation of predictive model equations that reveal the mathematical relationship between dependent variables and dependent response values.
- Ensures that results are displayed in three-dimensional or contour graphics.
- Allows simultaneous modification of the parameters that are active on the result during the experimental run.

In this section, a sample hydrometallurgy study is represented comparative by using two different response surface methods. For this purpose, determination of experimental conditions, examination of results, evaluation of statistical data, and optimization of results were studied with central composite design (CCD) and Box-Behnken Design (BBD) layout. Assuming that oxidized copper ore (CuO) will be leached in the presence of sulfuric acid (H_2SO_4). Suggested reaction is as follows and effective parameters are H_2SO_4 concentration, leaching temperature, and leaching time.

$$CuO + H_2SO_4 \rightarrow CuSO_4 + H_2O \tag{5}$$

Experimental design will be considered as only one response value evaluation as copper extraction. **Tables 1** and **2** show the interval of the examined parameters for CCD and BBD, respectively.

Factor	Unit	−α	−1	0	+1	+α
H_2SO_4	M	2.64	4	6	8	9.36
Temperature	°C	11.36	25	45	65	78.64
Time	Min	9.55	30	60	90	110.,45

Table 1. Interval of examined parameters for CCD.

Factor	Unit	−1	0	+1
H_2SO_4	M	4	6	8
Temperature	°C	25	45	65
Time	Min	30	60	90

Table 2. Interval of examined parameters for BBD.

As seen in **Tables 1** and **2**, CCD allows for experimental researching at extra points, which represents distance from center design points (alpha points). On the other hand, BBD has same design points with CCD except alpha values. Here, (−1), (0), and (+1) represent the design low value, design high level, and design center level, respectively. This fact points out that these alpha points give rotability to design layout. Also, interval of investigating parameters is (4–8 M) of H_2SO_4, (25–65°C) of leaching temperature, and (30–90 min) of leaching time. Assumed results are entered in the experimental conditions specified by the software. While CCD has 20 experimental runs and 6 center points, BBD has 17 experimental runs and 5 center points. In **Figure 7**, experimental design model and its investigating interval of each parameter are seen.

4.1. Analysis of comparative experimental results

The experimental results mentioned above were obtained. Standard error of design graphics is shown for CCD and BBD in **Figures 8** and **9**. According to error distribution results, different graphic trends exhibit for CCD and BBD, so that error values increase as it is approaching the low design points in CCD model.

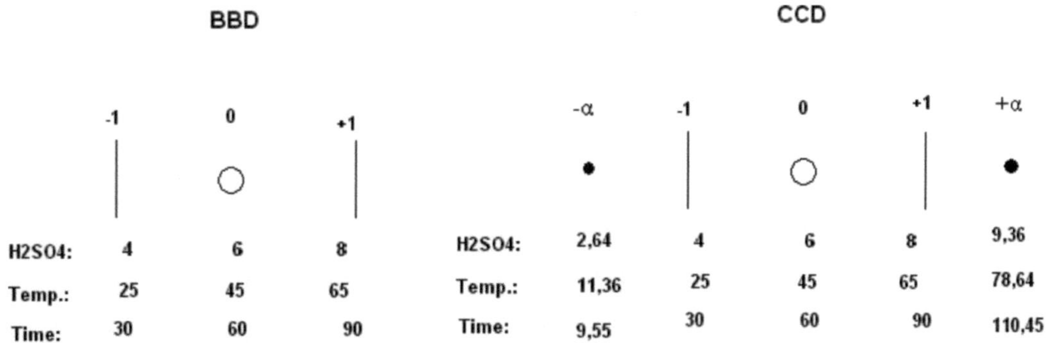

Figure 7. Experimental design for CCD and BBD.

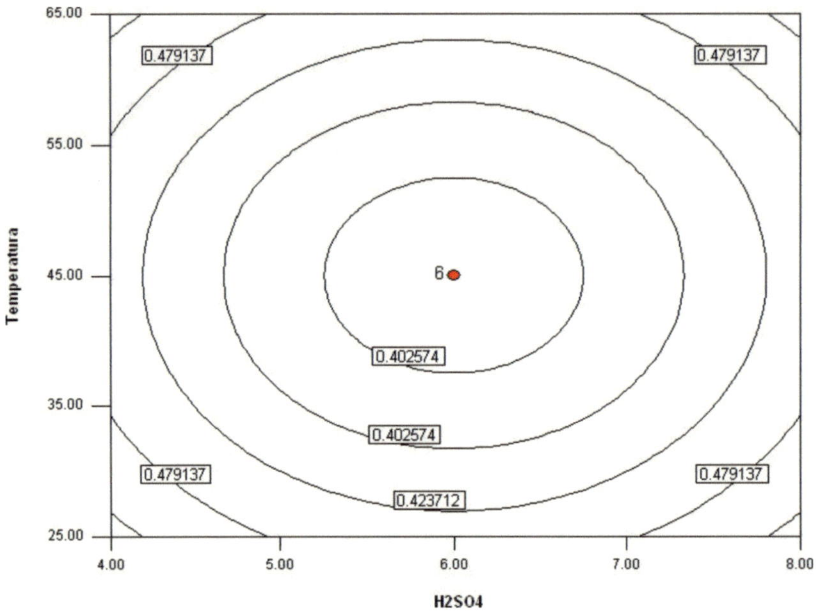

Figure 8. Standard error of central composite design model (Time: 60 min-constant).

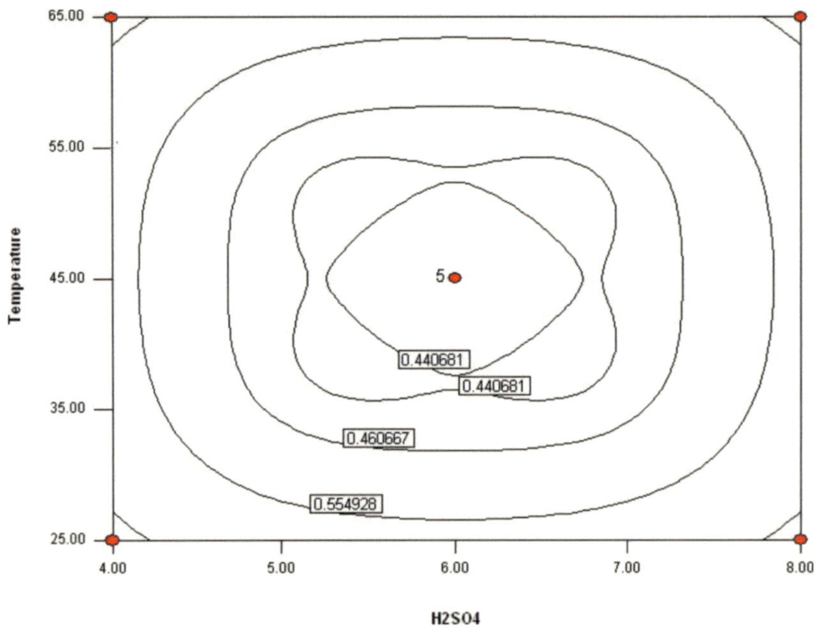

Figure 9. Standard error of Box-Behnken design model (Time: 60 min constant).

On the other hand, **Tables 3** and **4** show that sequential model sum of squares in which can be decided to fit what type source in design conditions. Practically, source can be chosen according to the high value of sum of squares. Thus, it can be said that appropriate source for model can be chosen as linear and quadratic sources for CCD and quadratic source for BBD.

http://dx.doi.org/10.5772/intechopen.71607

Source	Sum of squares	DF	Mean square	F value	Prob > F
Mean	58644.45	1	58644.45		
Linear	5950.16	3	1983.39	7.79	0.0020
2FI	543.37	3	181.12	0.67	0.5872
Quadratic	2324.65	3	774.88	6.42	0.0107
Cubic	1074.34	4	268.58	12.21	0.0048
Quartic	132.03	1	132.03	6.366E+007	<0.0001
Fifth	0.000	0			
Sixth	0.000	0			
Residual	0.000	5	0.000		
Total	68669.00	20	3433.45		

Table 3. Sequential model sum of squares for CCD [degrees of freedom (DF)].

Source	Sum of squares	DF	Mean square	F value	Prob > F
Mean	55461.24	1	55461.24		
Linear	620.75	3	206.92	1.72	0.2114
2FI	6.25	3	2.08	0.013	0.9977
Quadratic	1429.01	3	476.34	26.52	0.0003
Cubic	125.75	3	41.92	6.366E+007	< 0.0001
Quartic	0.000	0			
Fifth	0.000	0			
Sixth	0.000	0			
Residual	0.000	4	0.000		
Total	57643.00	17	3390.76		

Table 4. Sequential model sum of squares for BBD [degrees of freedom (DF)].

Analysis of variance (ANOVA) may be used to provide assistance with the interpretation of results. Essentially, ANOVA tables have numerous benefit data about model design such as interaction of parameters, lack of fit, degrees of freedom, etc. The ANOVA data of CCD and BBD are shown in **Tables 5** and **6**, respectively. In these tables, leaching parameters are symbolizing A, B, and C as sulfuric acid concentration, leaching temperature, and leaching time, respectively. Because of the quadratic of models, there are exponential and interaction between parameters too. On the other hand, statistical information about residual, lack of fit, and pure error for each model is available. ANOVA values also represent effective parameters on the result according to "F value" or "Prob > F" columns in which Prob > F value of parameters less than 0.05 are significant parameters on the result. Values greater than 0.1000 indicate the model terms are not significant.

In this case, A, B, A^2, and C^2 are significant model terms for central composite design while B, C, A^2, and B^2 are significant model terms for Box-Behnken design. The meaning of these significant models is that A and B parameters have linear effect on the results; also exponential

Source	Sum of squares	DF	Mean square	F value	Prob > F
Model	8818.19	9	979.80	8.12	0.0015
A	1734.48	1	1734.48	14.38	0.0035
B	3645.71	1	3645.71	30.22	0.0003
C	569.97	1	569.97	4.72	0.0548
A^2	1023.62	1	1023.62	8.49	0.0155
B^2	480.83	1	480.83	3.99	0.0738
C^2	1249.58	1	1249.58	10.36	0.0092
AB	36.13	1	36.13	0.30	0.5962
AC	231.12	1	231.12	1.92	0.1964
BC	276.12	1	276.12	2.29	0.1612
Residual	1206.36	10	120.64		
Lack of fit	1206.36	5	241.27		
Pure error	0.000	5	0.000		
Cor total	10024.55	19			

Table 5. ANOVA for response surface quadratic model of CCD.

Source	Sum of squares	DF	Mean square	F value	Prob > F
Model	2056.01	9	228.45	12.72	0.0015
A	55.13	1	55.13	3.07	0.1233
B	253.13	1	253.13	14.09	0.0071
C	312.50	1	312.50	17.40	0.0042
A^2	931.64	1	931.64	51.86	0.0002
B^2	331.64	1	331.64	18.46	0.0036
C^2	55.33	1	55.33	3.08	0.1227
AB	6.25	1	6.25	0.35	0.5738
AC	0.000	1	0.000	0.000	1.0000
BC	0.000	1	0.000	0.000	1.0000
Residual	125.75	7	17.96		
Lack of fit	125.75	3	41.92		
Pure error	0.000	4	0.000		
Cor total	2181.76	16			

Table 6. ANOVA for response surface quadratic model of BBD.

symbols (like A^2, C^2) have second-order effect on the results for central composite design model. For Box-Behnken design, B and C parameters have linear effect and A^2 and B^2 of parameters have exponential effect on the results. According to ANOVA of both models, interesting approaches are arising on the account of different parameters and their different exponential values are significant in each design model despite two models have been built in same experimental data.

Std. dev.	10.98	R-squared	0.8797
Mean	54.15	Adj R-squared	0.7714
C.V.	20.28	Pred R-squared	0.0431
PRESS	9592.29	Adeq precision	10.007

Table 7. Statistical data from CCD model.

Std. dev.	4.24	R-squared	0.9424
Mean	57.12	Adj R-squared	0.8683
C.V.	7.42	Pred R-squared	0.0778
PRESS	2012.00	Adeq precision	10.228

Table 8. Statistical data from BBD model.

This situation can only be explained by the presence of extra experimental points, so that it is seen how important the choice of design model in optimization. Some statistical data from both models are shown in the following (**Tables 7** and **8**). Important data may be sorted as "standard deviation (std. dev.)," R^2 "(R-squared)," "adjusted R^2 (adj R-squared)," and "adequate precision (adeq precision)." According to table data of CCD and BBD, standard deviation of CCD model is higher than BBD model that their values are 10.98 and 4.24, respectively. Statistically, model fit is better if standard deviation is low, so that this case influences the R-squared and adj R-squared values. In this context, R^2 and Adj-R^2 values of BBD model is higher than CCD model. Both type R^2 value is important for model fitting, in which the high one is preferred. When comparing between the R^2 values according to this result, it can be said that BBD model is more appropriate for the same experimental conditions. The probable cause of this case is that BBD has a more linear design model, but this linearity is valid only in conditions of parameters changing that have a linear effect on the response. On the other hand, adeq precision measures the signal to noise ratio. A ratio greater than 4 is desirable. Both model ratios (10.007 and 10.228) indicate an adequate signal. These models can be used to navigate the design space.

Result model equation can be created with RSM. As shown in Eqs. (6) and (7), final equation in terms of actual factors of CCD and BBD have different coefficient (the coefficients rounded to two digits).

$$\begin{aligned}\text{Cu \% (CCD model)} = {} & -82.59 + 23.15^{*}\, H_2SO_4 + 1.21^{*}\, \text{Temp} + 0.48^{*}\, \text{Time} - 2.11^{*}\, H_2SO_4^{\,2} \\ & - 0.01^{*}\, \text{Temp}^2 - 0.01^{*}\, \text{Time}^2 + 0.05^{*}\, H_2SO_4^{\,*}\, \text{Temp} \\ & + 0.09^{*}\, H_2SO_4^{\,*}\, \text{Time} + 9.79^{*}\, 10^{-3*}\, \text{Temp}^{*}\, \text{Time}\end{aligned} \tag{6}$$

$$\begin{aligned}\text{Cu \% (BBD model)} = {} & -164.77 + 47.34^{*}\, H_2SO_4 + 2.47^{*}\, \text{Temp} + 0.69^{*}\, \text{Time} - 3.72^{*}\, H_2SO_4^{\,2} \\ & - 0.02^{*}\, \text{Temp}^2 - 4.03^{*}\, 10^{-3*}\, \text{Time}^2 - 0.03^{*}\, H_2SO_4^{\,*}\, \text{Temp} \\ & - 2.29^{*}\, 10^{-16*}\, H_2SO_4^{\,*}\, \text{Time} - 6.57^{*}\, 10^{-18*}\, \text{Temp}^{*}\, \text{Time}\end{aligned} \tag{7}$$

The above equations are useful model equations. Especially, if a parameter data is entered except for the design points, this model equation gives possible outcome. As it can be remembered, design points are 4–8 M of H_2SO_4, 25–65°C of temperature, and 30–90 min of time. Let us run the model equation with any values outside the design points where it is wondered copper extraction result (response value) for 5 M of H_2SO_4, 30°C of temperature, and 80 min of time. According to the model equation result, it is obtained Cu%: 49.11 for central composite design model and Cu%: 59.94 for Box-Behnken design model. Although the statistical data indicate BBD, the most accurate result can be achieved by comparing the result of the experiments made under these conditions.

The residual plots were examined for the model adequacy for each metal extraction design values. In **Figures 10** and **11**, the normal percent probability-studentized residual plot and predicted-actual plots are shown for copper extraction of two different model designs. All the normal probability plots show how well the model satisfies the assumptions of the ANOVA where the studentized residuals measure the number of standard deviations separating the

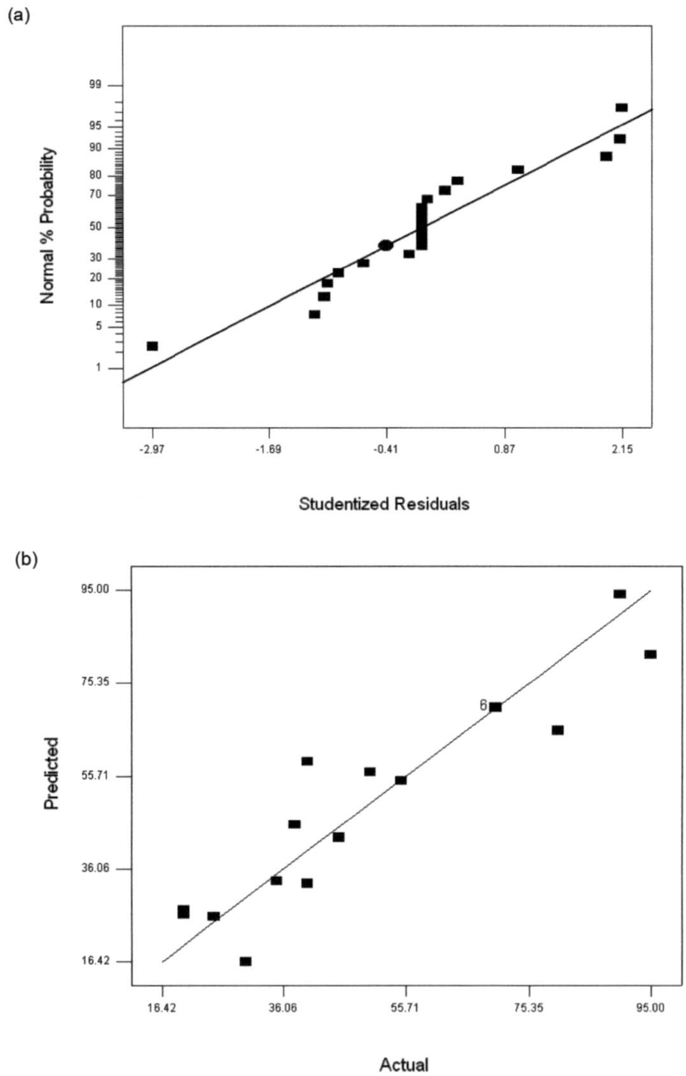

Figure 10. Diagnostic graphics of CCD model: (a) normal probability and (b) predicted vs. actual graphics.

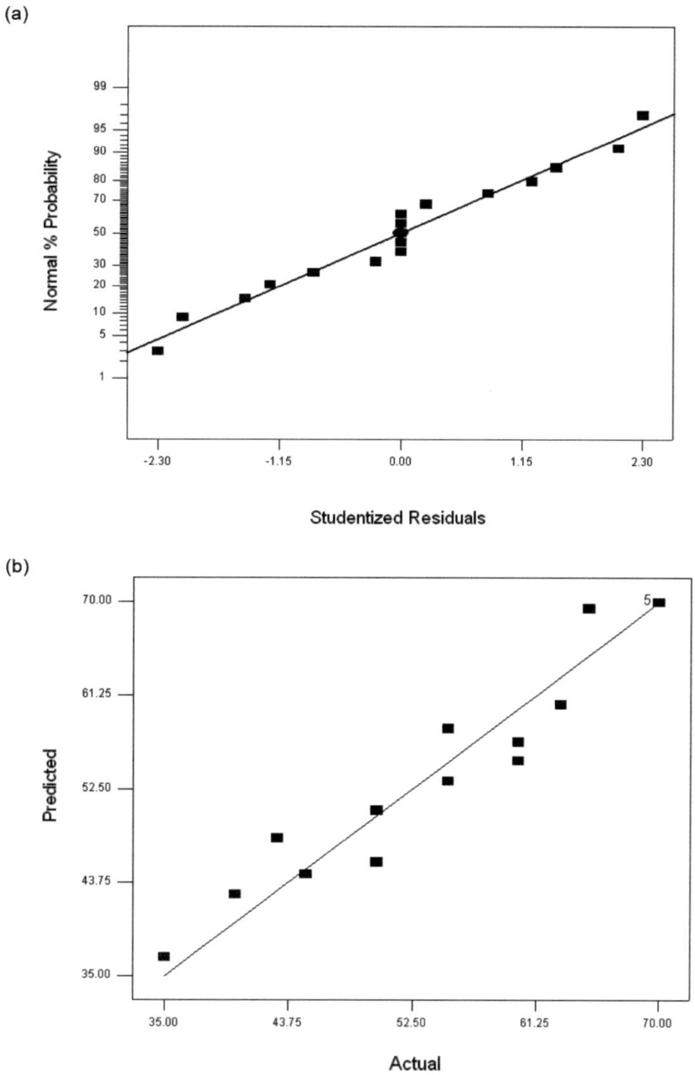

Figure 11. Diagnostic graphics of BBD model. (a) normal probability and (b) predicted vs. actual graphics.

actual and predicted values [19]. It is not desirable to have an S-shaped distribution in the normal distribution curve; however, it is desirable that the point distribution should be around line on the predicted-actual graphs. Last stage of design is to be the determination of optimization criteria. Optimization criteria of the both design models were held as the maximum copper extraction and three parameters interval was kept in range scale. Under these conditions, the software purpose solution points where different response values (Cu extraction) can be obtained by changing leaching parameters. These tables are shown in **Tables 9** and **10**.

As seen in **Tables 9** and **10**, desirability of BBD model is higher than CCD that it is indicated that the model adaptation for the current situation is better in BBD even though copper extraction values are higher in CCD model. The response surface graphs for metal extractions under the optimum condition are shown in **Figure 12** for CCD and **Figure 13** for BBD.

The presence of onion rings in the graphics indicates that optimum regions were obtained under these conditions. While Box-Behnken design model gives the optimum regions, it does not provide with central compost design model. This is a very controversial issue so that when considering the design model, all conditions must be taken into account.

These three-dimensional result graphics are very useful in terms of observing the results. In this kind of graphs, the effect of more than one parameter on the result can be determined simultaneously, and also how it affects the result by interaction between the changing parameters. **Figure 12** shows that high copper extraction can be obtained at high values of the parameters. For instance, 95% of copper extraction can be achieved under conditions: the leaching temperature of 65°C, leaching time of 90 min, and H_2SO_4 concentration of 8 M. An increase in extraction value with increasing values is not an acceptable approach in such studies. Because the goal is to achieve optimum conditions and is well-known that over consumption already provides high extraction values.

Number	H_2SO_4	Temp	Time	Cu	Desirability
1	8.00	65.00	88.55	94.1577	0.989
2	8.00	65.00	88.59	94.1574	0.989
3	8.00	65.00	88.74	94.1571	0.989
4	8.00	65.00	88.89	94.1561	0.989
5	8.00	65.00	88.87	94.1542	0.989
6	8.00	64.99	88.51	94.1542	0.989
7	8.00	65.00	89.70	94.1434	0.989
8	8.00	65.00	87.31	94.1423	0.989
9	7.94	65.00	90.00	94.0657	0.988
10	8.00	65.00	80.37	93.4687	0.980

Table 9. Optimum solution points of Cu percentage for CCD model.

Number	H_2SO_4	Temp	Time	Cu	Desirability
1	6.65	56.96	78.94	71.7234	1.000
2	6.14	51.34	65.60	72.0147	1.000
3	5.31	50.57	80.33	70.8928	1.000
4	5.53	48.11	79.95	71.8394	1.000
5	7.03	49.77	85.76	70.8148	1.000
6	5.88	53.32	61.76	70.9737	1.000
7	6.64	55.29	65.82	70.7235	1.000
8	6.56	54.94	65.38	71.0067	1.000
9	6.06	53.81	65.78	71.8761	1.000
10	5.76	47.27	60.14	70.046	1.000

Table 10. Optimum solution points of Cu percentage for BBD model.

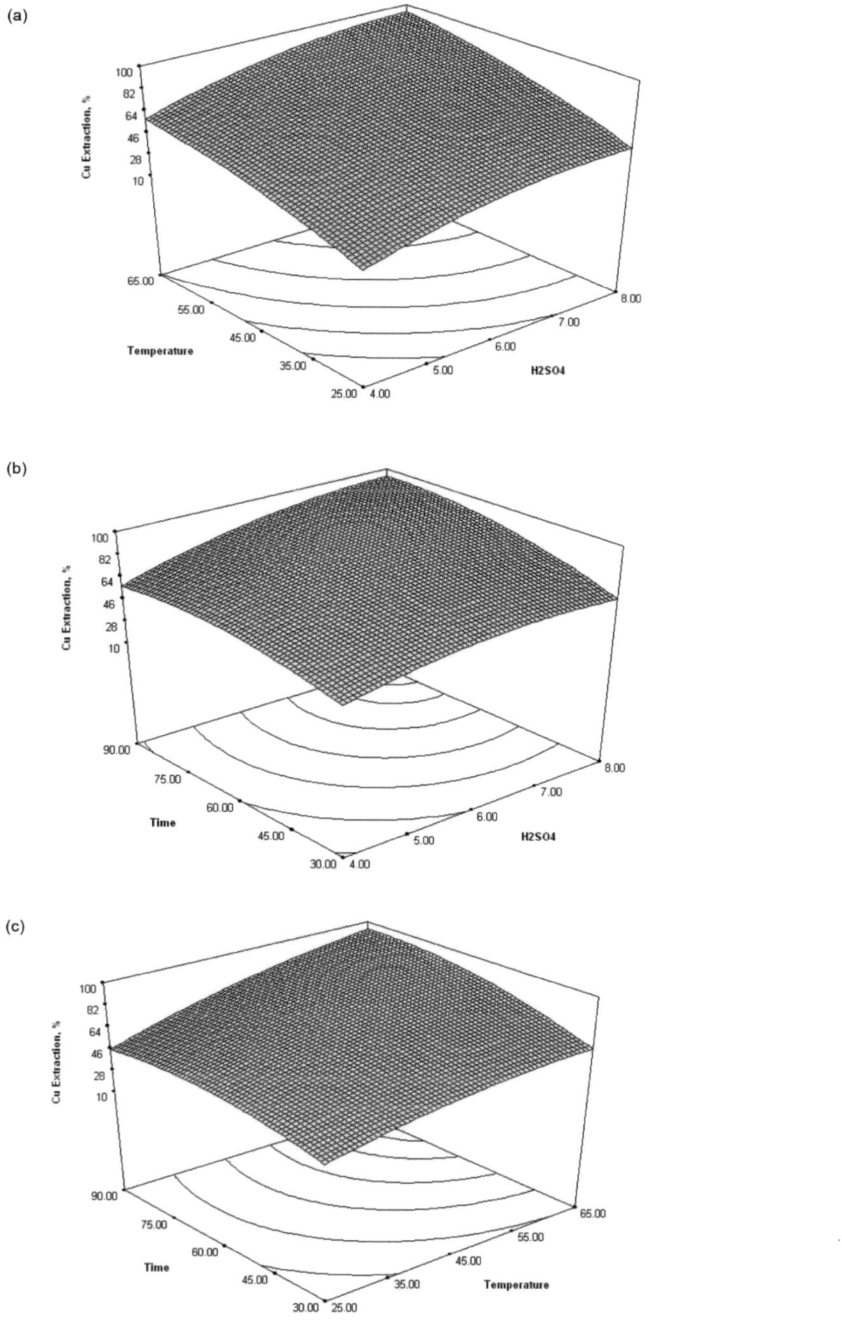

Figure 12. The effect of parameters interaction on the result for CCD model.

On the other hand, **Figure 13** shows optimal copper extraction under optimum conditions. It is clear that graphs of temperature-H_2SO_4 and time-H_2SO_4 interaction have completed circle lines similar to onion rings. The center region of these rings point out hump on the surface graphic (3D graphics), which indicates the optimum point. It is noticed that other sides of parameter values

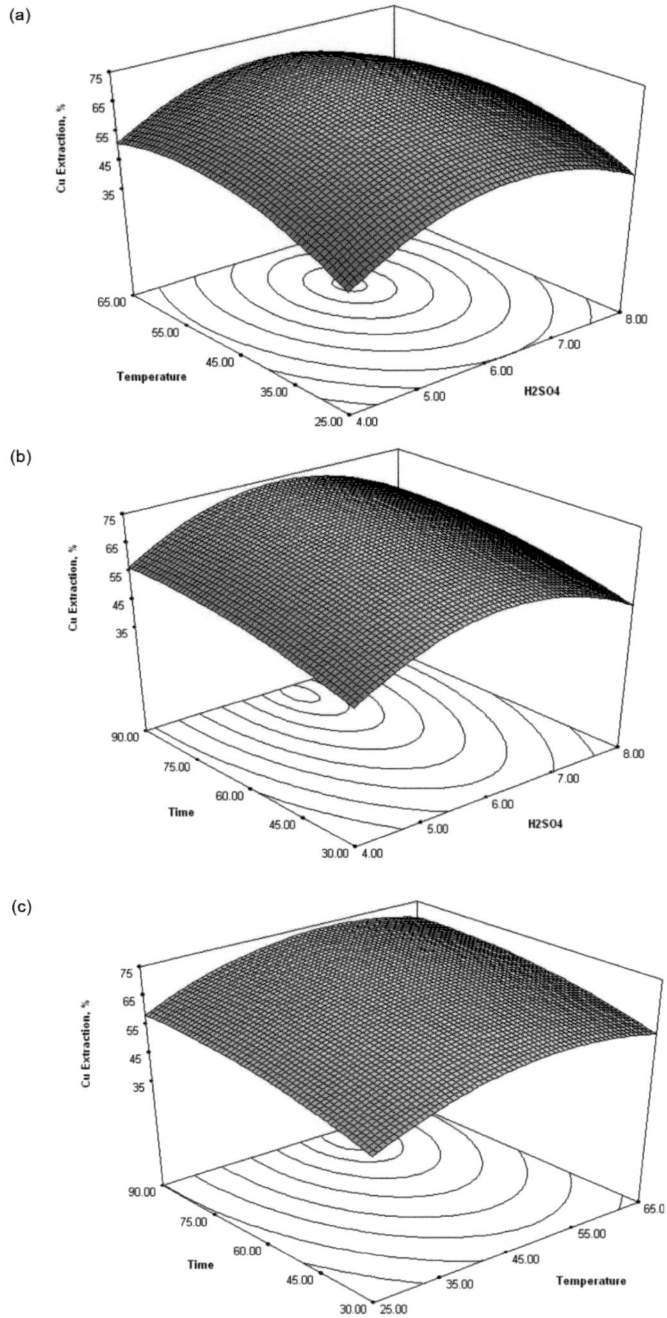

Figure 13. The effect of parameters interaction on the result for BBD model.

of hump (increasing and/or decreasing parameter values) cause decrease of copper extraction in which it is considered to be moving away from the optimum region. According to **Figure 13a**, approximately 70% of Cu extraction can be obtained in presence the 45–55°C of leaching temperature, and 5.5–6.5 M of H_2SO_4 concentration. Also, **Figure 13b** shows that optimum conditions of parameters are 75–90 min of leaching time and 5.5–6.5 M of H_2SO_4 concentration.

It is noted that if the parameter values are increased on the CCD for the same operation, the copper extraction is increased while the increased parameter values in the BBD cause the extraction values to drop significantly. This tendency arises from the examination of the effects of the parameters.

When two results about one factor plot are compared, copper extraction increases with increasing H_2SO_4-time-temperature for each plots of CCD model whereas copper extraction decreases or reaches the plateau with increasing parameters values after optimum condition points. Of course, it is not ignored that these factors are in an interaction.

In this chapter, presumed experimental data about hydrometallurgical copper extraction accompanied by three parameters were applied to two different design models, central composite design (CCD) and Box-Behnken design (BBD), in order to compare the results. As defined in the sections above, different statistical results and approaches were obtained although the same basic design values are entered as data. There are numerous design models, but the user should select the most suitable for own process. Namely, first of all we need to know the process well. As such, design models are not susceptible to unexpected changes in the process. According to the above model comparison, Box-Behnken design model is more appropriate for this process. It is implied that central composite design is not appropriate for this process, although it is generally more sensitive because of owing extra design points and rotatable feature. Although, only two models are compared here and different results are obtained, it is impossible to test all the models for a process. Instead of this, it is always more convenient to recognize model designs and processes and select a model accordingly.

Despite the optimization comparison above on a numerical example, similar study can be done for the modeling of nonnumerical (it will attribute as categorical factor) processes. In this context, the use of response surface methodology for optimization purposes of some science branches such as mineralization, geochemical, stirring profile of any process can be considered to increase productivity.

Author details

Mehmet Deniz Turan

Address all correspondence to: mdturan@firat.edu.tr

Engineering Faculty, Department of Metallurgical and Materials Engineering, Fırat University, Elazığ, Turkey

References

[1] Gunn G. Critical Metals Handbook, British Geological Survey. Nottingham, UK: John Wiley & Sons, Ltd; 2014. p. 669

[2] Han KN, Meng X. Recovery of copper from its sulphides and other sources using halogen reagents and oxidants. Minerals & Metallurgical Processing. 2003;**20**:160-164

[3] Kayadeniz İ, Sağdik U. Recovery of copper and cobalt by amonia leaching from Küre converter slag. Chimicia Acta Turcica. 1977;**5**:183-188

[4] Yang Z, Rui-lin M, Wang-dong N, Wang H. Selective leaching of base metals from copper smelter slag. Hydrometallurgy. 2010;**103**:25-29

[5] Li Y, Perederiy L, Papangelakis VG. Cleaning of waste smelter slags and recovery of valuable metals by pressure oxidative leaching. Journal of Hazardous Materials. 2008;**152**:607-615

[6] Yıldız N. 2014. Cevher Hazırlama ve Zenginleştirme (in Turkish). Erdem Publishing Company, Ankara, ISBN: 9789759677954(tk.no), p: 14

[7] Mead R, Pike DJ, Biometrics Invited A, Paper A. Rewiev of response surface methodology from a biometric viewpoint. Biometrics. 1975;**31**(4):803-851

[8] Walker CE, Parkhurst AM. Response surface analysis of bake-jab data with a personel computer. Cereal Foods World. 1984;**29**(10):662

[9] Joglekar AM, May AT. Product excellence trough design of experiments. Cereal Foods World. 1987;**32**(12):857-860

[10] Montgomery DC. Design and Analysis of Experiments. 5th ed. USA: John Wiley & Sons, Inc.; 2001. pp. 427-429. ISBN: 0-471-31649-0

[11] Design-Expert® Software. Version 6 User's Guide. Stat-Ease, Inc. Minneapolis, 2000 (software guide)

[12] Demir F, Dönmez B. Optimization of the dissolution of magnesite in citric acid solutions. International Journal of Mineral Processing. 2008;**87**:60-64

[13] Farahmand F, Moradkhani D, Safarzadeh MS, Rashchi F. Brine leaching of lead-bearing zinc plant residues: Process optimization using orthogonal array design methodology. Hydrometallurgy. 2009;**95**:316-324

[14] Khajeh M. Optimization of microwave-assisted extraction procedure for zinc and copper determination in food samples by Box-Behnken design. Journal of Food Composition and Analysis. 2009;**22**:343-346

[15] Mohapatra S, Pradhan N, Mohanty S, Sukla LB. Recovery of nickel from lateritic nickel ore using *Aspergillus niger* and optimization of parameters. Minerals Engineering. 2009;**22**:311-313

[16] Jiao H, Peng W, Zhao J, Xu C. Extraction performance of bisphenol A from aqueous solutions by emulsion liquid membrane using response surface methodology. Desalination. 2013;**16**:36-43

[17] Nasab EM, Sam A, Milani A. Determination of optimum process conditions for solvent extraction of thorium using Taguchi method. Journal of Radioanalytical and Nuclear Chemistry. 2011;**287**(1):239-245

[18] Nejad Hassan H, Shafiei Zadeh S, Alam S. Modelling of platinum extraction by Aliquat 336 utilising RSM technique. Canadian Metallurgical Quarterly. 2013;**6**:342-347

[19] Körbahtı BK, Rauf MA. Determination of optimum operating conditions of carmine decoloration by UV/H_2O_2 using response surface methodology. Journal of Hazardous Material. 2009;**161**:281-286

Chapter 9

Expected Return on Capital in Mining Industry

Aneta Michalak

Additional information is available at the end of the chapter

http://dx.doi.org/10.5772/intechopen.72689

Abstract

Capital is a necessary element of each economic activity. In the enterprises functioning in capital-consuming industries, such as mining industry, the problem of capital becomes more complex and is followed by a number of problems. Investors (both owners and creditors) expect the return on invested capital, taking into consideration the risk level connected with the activity that is to be financed by them. The problem raised in this work is related to the determination of the ways of calculation of the expected return on capital from the point of view of capital provider, with the inclusion of the specificity of mining industry. The universal calculation methods in use are difficult to be applied in the enterprises from mining industry because of the specific character of risk emerging in these enterprises, unique character, and high capital consumption. The author suggests modification of one of the most popular methods and presents her own, new solutions in this area, adjusted to the specificity of mining enterprises. The solutions presented allow a more realistic look on the issue of financing mining activity.

Keywords: cost of capital, mining industry, mining enterprises, risk factors, return on capital

1. Introduction

Capital is a necessary element of each economic activity. Its significance is increasing along with the increase of capital consumption in the activity. In the enterprises functioning in capital-consuming industries, such as mining industry, the problem of capital becomes more complex, which is followed by a number of problems that do not emerge in the enterprises functioning in other industries. Taking the size of capital used in mining industry into account and the problems linked to its management, the measurement of the return on capital expected by investors became the objective of this research. The problem of how to calculate the expected return rate is known, both in scientific circles and business practice. However, the universal calculation methods in use are difficult to be applied in the enterprises from mining industry due to a specific character of risk occurring in such enterprises, unique type of activity, and high capital consumption. In order to meet such a goal, diagnostic and conceptual research was conducted. As a part of diagnostic research, the method, often used, of estimating the return rate on capital called capital asset pricing model (CAPM) was analyzed and verified, but that does not fully resolve the problem. Therefore, it was used as the grounds for conceptual research, as a part of which an attempt was made to design a new method including the conditions of mining industry. The suggested solutions, to a great extent, facilitate the process of calculation of the expected return on capital used to finance mining activity; even in cases when such calculation was not possible or did not provide rational results, they make this process possible and give measurable results. The solutions presented allow a more realistic look on the issue of financing mining activity. They may find use in the efficiency calculus of mining enterprises, value pricing of mining enterprises, in the area of development of optimal structure of financing sources, when obtaining capital through shares or bonds issuance, etc.

2. Essential definition set

In order to define the expected return rate on capital, firstly, one should refer to the definition of capital. Capital is one of the most often used notions in economic sciences and many others. At the same time, it is interpreted in various ways, both in theory and economic practice. For this reason, some kind of theoretical pluralism appeared, where the proper essence is sought for the particular area of knowledge. For example, one may talk about real capital, tangible, operating, money, human, intellectual, social capital, etc.

The definition of capital started to be used in the Middle Ages, although there is a proof that the foundation for the conception of capital has already appeared in prehistoric times, several thousands of years BC. The conceptions linking capital with a corresponding return rate have been developed as the definition of capital was shaping. Capital was initially defined as the monetary value being able to earn interest [1]. The notion of capital also referred to all the goods possessed [2], and as the time went by, it encompassed a wider range of possessions, going beyond a traditional understanding of assets and including other, even intangible values. At present "capital" is a wide and equivocal category. Various ways of approach to defining and interpreting this definition appear in both theory and economic practice. Capital is perceived as strength, which is the basis for progress; it increases work efficiency and creates the wealth of nations [3]. Capital is at the same time noticed as a category responding to goods (resources) possessed, used to run and develop economic activity (tangible or resource conceptions) or responding to savings gathered (monetary conceptions). Its basic feature is the ability to grow (see more in [4]). Such growth may take a form of income, financial revenue (percentage), value added, and others. However, a fact is emphasized that the growth occurring as a result of capital use takes place after a particular amount of time (compare with [5]). These are only some of the conceptions of capital appearing in the history.

In the process of enterprise management, capital may also be analyzed in the forms of existence as **real** capital and **financial** capital. From this perspective, real capital refers to the particular material goods participating in the production process. It is identified with a factor of production, which subjectively materializes on each stage of the particular action process, emerging at the same time from the combination of other elements of the production process, material resources, work, and time, which are combined in an entrepreneurial performance of people [6]. Accordingly, financial capital does not possess a form of resources used to produce goods and services; however, it constitutes the source of financing for real capital [7]. This way of capital perception appears in **corporate finance** too. In that area the streaming character of capital is marked, which comes from the fact that capital equals to the stream introduced to the enterprise and circulates in its operations, generating value increase [6]. The dual character of real and financial capital may be found in **accounting**. The definition of capital is treated dually as capital materialized in a form of assets, serving an active role, and as an abstract capital, having its reflection in enterprise's liabilities, playing a passive role [8]. Following this pattern, capital is defined as funds entrusted to the enterprise by its owners and creditors [9]. Such perception of capital is used in this work.

Despite a great variety of approaches to defining capital, it should be noticed that in most conceptions of capital there is one inherent feature, namely, the ability to generate profit. Some people identify this profit with income (most often in asset, production approach to capital) [10]. However, the most common understanding of capital profits is interest, which mainly relates to capital in value depiction, that is, money. In this work it is assumed that the return on invested capital consists of the **return rate on capital**, expected by investors (both owners and creditors) at a given risk level [9]. It is connected with opportunity cost, that is, expected return rate by the investors, which they may resign from when choosing a particular variant of action and therefore resigning from other possibilities available at that moment [11].

The significance of risk should be emphasized when defining the expected return rate. Investments of the same risk level have the same return rate. Consequently, the return rate expected by investors, in case of investing capital in a particular undertaking, should be at least equal to the highest total return that the investors could expect if they invested in an alternative portfolio of securities with a comparable risk [12]. According to this approach, the more risky the enterprise's activity is, the higher return rate should be generated (compare with [13]). The expected return rate defined in this way has its interpretation in the price that the enterprise pays for the possibility of using capital. This price is expressed in a form of interest rate, reflecting the relation of expenses borne by the capital provider in a yearly scale due to putting capital at someone's disposal and the value of this capital. In this depiction the expected return rate is strictly connected with the source of capital's origin and accompanies capital coming from any source. It may be described as interest, dividends, other benefits, or even opportunity cost for capital providers. In subject literature the division emerged into the expected return rate on equity and debt capital. Individual financing sources, being the components of these groups, are characterized by a different return rate for capital providers. Usually, the return rate even on the same type of capital is different. For example, a different interest rate is assigned to the investment loan in different banks. It results from the contents of particular agreements.

The expected return rate may be understood in a different way from the point of view of capital provider and recipient. For capital recipient the desired return rate at a given risk level is the cost of capital. However, some circumstances should be taken into account, when the return rate may be higher or lower than the cost of capital. For capital recipient the cost of capital is lower than the return rate expected by capital provider, when tax benefits emerge thanks to decreasing the taxable amount by the interest on capital. It consists in the fact that the interest paid by the enterprise decreases gross financial result, lowering at the same time taxable amount. This phenomenon is called **tax shield** and causes that the total cost of debt is lower than the return rate expected by creditors. Also, an adverse situation may take place when the cost of capital for capital recipient is higher than the return rate for capital provider. It is connected with bearing additional costs by capital recipient. For example, when the enterprise gains capital through securities issuance, the cost of capital is higher than the return rate expected by investors, due to the costs related to securities issuance (compare with [14]).

3. Risk as the basic indicator of expected return rate

Risk is considered to be one of the most important indicators of the expected return rate, especially in mining industry, due to the capital-consuming character of this industry and long period of return for the outlays put. **Risk premium**, which should compensate risk borne by capital providers, constitutes an additional remuneration for capital providers over the return rate on bonds. Risk connected with a particular enterprise may be divided into specific risk (also called individual risk) and systematic risk (also called market risk). Both categories of risk mean a probability of occurrence of future incidents affecting the results of enterprise's activities; however, **specific risk** refers to the incidents that may be partially controlled by the enterprise and generally come from its internal decisions. The source of specific risk may be the way of enterprise management, financial liquidity, level of financial and operating leverage, availability of resources, and other factors. The measurement of this risk is disseminated neither in literature nor in practice. In general, ready-made rankings are used, or it is treated as an average risk of enterprises from the same industry. In an international scale, the rankings of industry risk are examined by, among others, A. Damodaran. Nevertheless, it does not fully correspond to specific risk as the rankings do not include the individual risk generated inside the enterprise. The need to include this risk is noticed in the subject literature, but there are no model solutions showing how to accomplish that. Usually, the component related to specific risk is indicated in an arbitral way, or based on experts' opinions. However, **systematic risk** refers to the general conditions of management; it is a result of the impact of external forces, out of the subject's control when exposed to that risk. It is connected with macroeconomic factors, such as economic situation, monetary and tax policy, social situation, legal system, etc. In connection with the occurrence of specific and systematic risk, the expected return rate on capital should include specific risk premium and systematic risk premium. Below, the solution to that problem is suggested, using appropriate indicators.

Risk generated inside the enterprise may be generally divided into operational and financial risk. **Operational risk** is most often perceived as a probability of operating loss or of not achieving the expected level of operating profit coming from inadequate or faulty internal processes, people, and systems or also from external incidents [15]. In a similar context, risk is connected with uncertainty toward the level of future cash flows of the enterprise [9]. Operational risk is strictly connected with the enterprise, industry, and close environment's specificity that the enterprise functions in. It is dependent from the complexity of enterprise as an organization, systems used, production process, products, or services (compare with [16]). In particular, operational risk depends on relationship between operating costs and sales revenues and on cost structure, including fixed and variable costs. The share of fixed cost in total operating cost determines the reaction of operating income (EBIT—earnings before interest and taxes) to change in sales revenues. This relation is the basis of the mechanism of **operating leverage**, which is considered to be the basic measure of enterprise's operational risk [17]. When the level of fixed cost in the structure of operating cost is high, also the level of operating leverage is high, and at the same time, risk of changes in sales revenues increases (if fixed cost is too high, even small change of price or sales amount may significantly influence the level of operating income). Operational risk is high in such situation. The degree of operating leverage (DOL) may be calculated as a function of EBIT change, and sales revenues change using information about the value of fixed cost in the following way [18]:

$$DOL = \frac{\frac{\Delta EBIT}{EBIT_0}}{\frac{\Delta S}{S_0}} = \frac{S_0 - Cv}{EBIT_0} \tag{1}$$

where $EBIT_0$ is the earnings before deducting interest and taxes, $\Delta EBIT$ is the change in EBIT, S_0 is the sales revenues, ΔS is the change in sales revenues, and Cv is the variable operating cost.

Financial risk is connected with the way of financing enterprise's activity, and the basic factor determining its level is capital structure [19]. Financial risk increases along with the growth of debt. It is suggested to parameterize financial risk using a substantially selected set of diagnostic indicators. When choosing them, firstly the attention was paid to the fact that financial risk is connected with the level of financial leverage in the enterprise. **Financial leverage** is the basic indicator reflecting capital structure in the enterprise. Its level expresses the relation of debt capital to equity, which may be written as

$$\textit{financial leverage ratio} = \mathrm{D/E} \tag{2}$$

where D is the debt burdened with interest and E is the equity.

If the financial leverage ratio in the enterprise is higher than 0 (only when the enterprise is financed by debt capital), the effect of financial leverage appears. It consists in a situation when, due to financial costs resulting from interest on debt, possible fluctuations of earnings per share are more than proportional to fluctuation of operating income. The higher the financial cost, the stronger the effect of financial leverage. It may be positive when it translates into increase of earnings per share or negative when such earnings decrease. The degree of this effect depends on the share of fixed financial cost in total cost of the enterprise [20].

The effect of financial leverage is measured using the **degree of financial leverage** (DFL). It is a measure of the assessment of enterprise's financial risk that informs by how many percent earnings per share change (rise or fall) due to 1% change (rise or fall) in operating income. It is expressed through a relation of relative change of earnings per share to a relative change of operating income, which may be shown as [21]

$$DFL = \frac{\Delta EPS/EPS}{\Delta EBIT/EBIT} = \frac{EBIT}{EBIT - I} \quad (3)$$

where EBIT is the earnings before deducting interest and taxes, EPS is the earnings per share, $EPS = \frac{(EBIT - I) \times (1 - I)}{n}$, I is the interest as fixed cost of debt capital (financial cost), and n is the number of issued shares.

The indicator of the degree of financial leverage is calculated for the particular value of EBIT. If EBIT is close to the cost of servicing debt, which means net income is close to zero, then the level of DFL is high. Along with EBIT rise, the level of DFL decreases (the share of fixed cost of servicing debt falls in relation with operating income), which proves that financial risk decreases. The higher the share of debt in the enterprise's capital structure and the higher the cost of servicing debt representing the return rate expected by capital providers, the higher the effect of financial leverage and the higher the value of DFL. When the enterprise is financed by equity only (I = 0), the effect of financial leverage does not occur and DFL = 1. The effects of financial leverage are different for various variants of corporate capital structure, however, the effects (positive or negative) are stronger when the share of debt capital increases in financing structure and the expected return rate on capital is higher.

In practice, based on DFL indicator expressing the relation of operating income and cost of debt servicing, ratings are assigned for enterprise's debt. Then, a corresponding financial risk premium resulting from debt is determined. The indicator showing the ability of servicing debt is used here—times interest earned (TIE) (used by renowned rating agencies, such as Standard & Poor's and Moody's Investors Service)—which measures the relation of operating income and cost of servicing debt capital [22]:

$$TIE = EBIT/I \quad (4)$$

TIE indicator is helpful in determining the border for debt capital employed. Low (close to 1) level of this indicator is a sign that the cost of servicing debt in the enterprise constitutes too much burden for operating income (the probability of insolvency increases). The fall under 1 of this indicator's value means that the enterprise loses the ability to pay interest on debt capital (operating income is not sufficient for covering financial costs) (compare with [23]).

Financial risk in the enterprise may be additionally pictured in practice using other indicators characterizing the financial structure of the enterprise. These include the share of equity in total capital (**equity to capital ratio**) and the share of debt in total capital (**debt to capital ratio**) (it is assumed that total capital is a sum of equity and debt capital with interest):

$$\text{equity} - \text{to} - \text{capital ratio} = \text{E}/\text{C}^{*}\,100 \quad (5)$$

$$\text{debt} - \text{to} - \text{capital ratio} = \text{D}/\text{C}^{*}\,100 \quad (6)$$

The first ratio informs about the level at which the enterprise is able to finance itself in the activity conducted. A high value of this ratio means solid financial grounds of the enterprise and ensures the creditors that in case of failure, the enterprise will be able to settle the debt incurred using equity. Debt to capital ratio informs about the level of enterprise's debt, that is, what part of corporate capital is debt [24].

Furthermore, when examining the corporate capital structure, also the level of assets coverage by equity (**equity to assets ratio**) is used. It allows determining if the enterprise fulfills the balance sheet's golden rule. This ratio is calculated in the following way:

$$equity - to - assets - ratio = E/A^{*}100 \tag{7}$$

When the value of this ratio is equal to or more than 100%, it means that the balance sheet's golden rule has been fulfilled. It says that fixed assets, which are engaged in conducting economic activity for a longer time, should be fully financed by equity as they are at enterprise's disposal in a long term.

The next variable considered to be an important indicator of financial risk is the level of net **operating capital (OC)**, also called working capital. It may be calculated as the subtraction of fixed liabilities (equity and long-term liabilities) and fixed assets. Net operating capital may have a positive, zero, or negative value. The first and the second situation may be considered as appropriate, but the fact of financing fixed assets by liabilities of shorter deadline than a year may be connected with the risk of losing financial liquidity. This happens in a situation of inability to settle due liabilities using current assets [25]. It is a particularly risky situation in case of capital-consuming industries such as mining industry, where we deal with great immobilization of assets of high value, which are very difficult to cash in most cases (underground buildings and objects, longwalls, professional mining devices, etc.).

An important problem during analysis of this area of the enterprise is determination of the limit of debt capital engagement. It may be identified using dependency analysis between the return on equity (ROE) indicator and return on net operating assets (RNOA) indicator. ROE informs about the value of net profit ascribed to one unit of employed equity. This indicator alone may serve as an estimator of specific risk premium. Nevertheless, it is not a perfect measurement method as it only bases on book value. The current financial reports do not include intellectual capital of the enterprise, which may lead to extremely high values of this ratio caused by decreasing the actual value of denominator in the formula [26]. **Return on equity** is indicated according to the formula

$$Return\ on\ equity\ (ROE) = net\ income/shareholder\ equity \tag{8}$$

Theoretically, the higher the value of this indicator, the more favorable the corporate financial situation. However, in each case, ROE indicator should be confronted with other measures of financial risk, for example, **RNOA**. In the further part of the analysis, ROE is related to RNOA. The latter one is indicated in the following way:

$$Return\ on\ operating\ assets\ (RNOA) = (NOPAT)/(NOA) \tag{9}$$

where T is the income tax, NOA is the *net operating assets*: NOA = fixed assets + current assets – short-term liabilities – operating notes (short term, without interest), and NOPAT is the *net operating profit after taxes*: NOPAT = EBIT*(1 – T).

Positive assessment is noted for the relation

$$ROE > RNOA \tag{10}$$

Systematic risk is interpreted as part of the risk connected with the enterprise's activity, which depends on a general situation on the market. It is the risk determined by external forces (general economy, force majeure), out of corporate control. A classic measure of systematic risk is **β coefficient** (also called a measure of a security or risk index), included in many calculation methods regarding the expected rate of return. β coefficient measures the volatility of a given security on the whole market, determining the level of market risk connected with investing in assets of a certain enterprise [27]. β coefficient may be defined as a relation of covariance of a security with market portfolio and variance of market portfolio. Covariance indicates the correlation between the volatility of return on investment in a given security and the volatility of return on investment in market portfolio. Furthermore, variance of market portfolio is a measure of systematic risk of this portfolio. β coefficient may be directly indicated from regression of past data regarding the particular instrument and market portfolio (stock exchange index). At that time β coefficient is a gradient of regression function (compare with [28]). β coefficient of value more than one is specific for a security with greater volatility of the return rate than the market index (aggressive securities); when it is lower than 1, it characterizes securities of lower volatility of the return rate compared to the return rate on market portfolio (defensive securities) (compare and see [29, 30]). In some cases β coefficient may be negative, which means that the return rate on a security reacts differently to changes than the return rate on market portfolio. Another untypical case is β equal to zero. In such situation the return rate on a security does not react to market changes (risk-free security, bonds in particular). In general, it is assumed that the enterprises with a higher level of β are more risky than the ones with lower β [31].

In practice, the value of β coefficient may be calculated according to the formula [11]

$$\beta = \frac{\operatorname{cov}(r_{it}, r_{mt})}{\operatorname{var}(r_{mt})} = \frac{\sum_{t=1}^{n}(r_{mt} - \bar{r}_m) \times (r_{it} - \bar{r}_i)}{\sum_{t=1}^{n}(r_{mt} - \bar{r}_m)^2} \tag{11}$$

where β is the beta coefficient, $\operatorname{cov}(r_{it}, r_{mt})$ is the covariance of market return with stock return, $\operatorname{var}(r_{mt})$ is the variance of market return, r_{it} is the return on stock in t period, r_{mt} is the return on market portfolio in t period, $\bar{r}_m$ is the average return on market portfolio in t period, $\bar{r}_i$ is the average return on stock in t period, and t is the period used as the basis for model parameters.

The calculation of β coefficient is connected with numerous problems that need to be resolved. One of the most important issues is to indicate the duration of estimation period, so-called estimation window. It is assumed at that time that based on daily return rates a shorter estimation window may be chosen (e.g., 2 years) and based on monthly return rates this period should be accordingly extended to several years [32]. In this area it is important to comply with the requirement of normal distribution of return rates on stock and benchmark index [33].

4. Analysis of limitations and possibilities of CAPM as the most popular method of estimating the expected return rate

In practice the estimation of the expected return rate is most often based on CAPM method. CAPM is a part of greater theory called *capital market theory* (CMT) [34, 35]. The basic assumption of CAPM is that a part of risk premium on the expected return on investment in securities is a function of market risk of these securities [36]. The expected return rate, in this approach, is a function of individual risk index, describing the volatility of return on stock of a given enterprise in relation with the return on the whole economy (usually represented by a particular market index) [37]. In terms of the analysis of market relationships, which lead to a certain investor behavior, CAPM refers to portfolio theory [38]. It assumes that the return rates on stock depend on the influence of market factor (stock exchange). It is connected with the regularity observed that on most developed stock exchanges, the fall or rise, measured by the change in the main market index, is accompanied by fall or rise of most stock prices. This basic assumption cannot be directly transferred to the expected return rate estimation in mining industry. The observation of capital market proves that stock of enterprises from this industry often reacts in an adverse way than stock of other enterprises. The conditions of mining activity are different to a great extent from the conditions of other economic activities. Nevertheless, for various reasons, CAPM is one of the most popular ways to calculate the expected return rate, also in mining industry. The usefulness of CAPM is noticed by renown authors of many American academic books—E.F. Brigham and L.C. Gapenski. These authors assume that CAPM bases on several, but not very realistic assumptions, and the model cannot be empirically verified. However, its logics provide the grounds for its application in the estimation process of expected return rate [39]. Similarly, A. Buckley, summarizing the dispute concerning CAPM, claims that even if empirical research does not fully confirm the usefulness of this method, it still is a convenient conception explaining the dependency of income and risk [40]. Despite the passing of time and many researches conducted on the problem of CAPM, it remains at a beginning stage of research, and, as the authorities in this field claim, explicit solutions are far to reach (compare with [41]). The difficulties in determining the expected return rate occur on the majority of developed markets of the world and are especially intensified on so-called emerging economies [42] that include the countries of Central and Eastern Europe, South America, and many countries of Southeast Asia. The difficulties particularly appear in specific industries, such as mining industry.

In general, there is no agreement in literature concerning the type of explanatory variables or the ways of determining the basic parameters, etc. Nevertheless, CAPM constitutes the basis for further attempts of searching for an optimal method of estimating the expected return rate. In most methods the attention is paid to risk pricing; however, only systematic risk is usually included, omitting specific risk of the enterprise. Within the frames of the actions taken in this area, it is necessary to add the specific features of the analyzed sector.

Capital asset pricing model assumes that a part of risk premium for the expected return on investment in securities is a function of market risk of these securities (more in [36]). The expected return rate in this approach is a function of individual risk index, describing the volatility of return on stock of the particular enterprise in relation with return on the whole economy (usually represented by a certain market index) [34, 35]. Return on capital invested in stock of the particular enterprise is connected with the market, which may be presented in the following formula:

$$Re = r_F + \beta \times (r_M - r_F) \tag{12}$$

where Re is the expected return on equity, r_F is the theoretical risk-free return rate, r_M is the average return rate on the market, and β is the beta coefficient determining the level of systematic risk (measure of dependence between the return rate on equity in the particular enterprise and the return rate on the market).

Risk-free return rate (r_F) is mainly indicated based on return on government securities [43]. In literature the return rate on long-term bonds or treasury bills is usually used [44]. **β coefficient** defines the level of systematic risk connected with investing in assets of the particular enterprise [27]. In turn, **average return on the market (r_M)** is most often adopted as the return rate on the basic stock exchange index. The analysis of risk premium bases on real returns from the past.

In some situation the assessment of market risk premium is especially difficult, for example, on emerging markets, where capital market does not provide sufficient data (not enough data or they are characterized by too high volatility, and they cannot be a source of reliable estimation of risk premium). Thus, in order to avoid complications in terms of r_M determination, total difference is calculated between the return rate on the market portfolio and risk-free return rate ($r_M - r_F$), which is called as **market risk premium (MRP)**. Next, it is assumed that this return rate is adequate for all market undertakings, as it is determined by objective macroeconomic conditions [45]. A disadvantage of the conceptions basing on the average return rate on the market is the unified expected return rate that does not include differences concerning risk between the enterprises performing on the same market. Estimation of the return rate is burdened with a mistake of averaging.

5. Alternative conceptions of expected return rate calculation: literature review

A response to CAPM limitations mentioned in the previous part can be alternative estimation methods of expected return rate. These include dividend growth model (DGM), Fama–French model, and arbitrage pricing model (APM).

DGM was created by J.B. Williams [46] and improved by M.J. Gordon and E. Shapiro [47]. The basic DGM bases on a formula describing the value of equity:

$$V_E = \sum_{t=1}^{n} \frac{DIV_t}{(1 + r_w)^t} \tag{13}$$

where V_E is the equity market value (stock market price x number of stock), DIV_t is the dividend paid in t period, where t = 1,2,…n, and r_w is the expected return rate on equity.

Such construction in a simplified way brings the estimation of equity down to examining two factors: expected dividend growth in the future and return rate that reflects systematic risk of the growth. However, it does not contradict the portfolio theory, which assumes that the return rate consists of dividend and capital profit or loss per share.

Dividend growth model has not found many supporters in literature or practice. The use of this model in its basic form, that is, with the assumption that the dividend growth is unlimited in time, requires developing the forecast of dividend growth from present up to unlimited yearly periods, which is very time-consuming. The difficulty of this model may also come from the necessity of determining the dividend level and its growth rate, which limits its application only to the enterprises characterized by stable growth and dividend policy aimed at regular payoffs, proportionally to the corporate profits. Mining enterprises do not comply with such requirement. Most of them do not have a policy of regular dividend payoff due to huge capital needs, connected with investment.

The model that constitutes a critical response to the standard version of CAPM is Fama and French three-factor model (Fama-French model), also called F-F model. Some researchers place this model among the nonstandard variations of CAPM. The authors of this model (E. Fama and K. French) proved in 1992 that the standard CAPM does not sufficiently explain the level of return rates on securities. Based on research conducted, they positively verified that the relationship between β and average return rate, analyzed in the years 1941–1990, is very weak and has a tendency to diminish. At the same time, they questioned the existence of positive relationship between the return rates on stock and estimated risk index for them (β). Nonetheless, they did not exclude the existence of covariance with the market. The factor, which reflects the volatility of securities to changes on the market, was supplemented with two other factors describing the return rates better. These were market value (market capitalization) and book-to-market ratio, describing financial risk. The supporters of this model claim that the factors included reflect the additional, undiversified (macroeconomic) risk, not taken into account in β coefficient in the standard CAPM version [48]. It is particularly important in case of subjects with low capitalization. Other voices, emerging in the discussion about the usefulness of Fama–French model, undermine the conclusions drawn by its authors. These are, among other, publications by S.P. Kothari, J. Shanken, and R.G. Sloan [49]. They claim that the yearly return rates include remuneration for risk, measured using β in the standard model. Furthermore, the model by E. Fama and K. French is criticized by F. Black, L.K. Chan, and J. Lakonishok, who state that those authors did not manage to prove to a sufficient degree that β from CAPM does not explain the expected return rates [50, 51].

APM is often called APT model (*arbitrage pricing theory*) [52] as it bases on arbitrage pricing theory. APM may be treated as multifactor version of CAPM (the nature of CAPM is single regression, APM is a multi-regression model) [53]. APM uses a set of macroeconomic factors determining the risk of a given business. This set is varied and may be developed if needed. Among the most often included factors, the following ones are mentioned: difference between long-term and short-term treasury bills or market rates, inflation rate, sales growth rate in production sector, exchange rates, or changes in forecast of macroeconomic variables such as GDP [37]. Also, substitute factors can be used if data about the primary factor are not available in the research period [54], which shows great flexibility of APM. One of the risk factors in APM may be the traditional β used in CAPM. In practice, other factors may be used, specifically for the particular activity, e.g., resources price index, energy price index, income level of certain groups of people, etc. [55]. APM does not have a standard character. Taking this into account, APM may be considered as an attractive estimation method of expected return rate; however, it raises serious problems when used in practice. Numerous attempts of return rate estimation using this model so far have not led to its dissemination among practitioners [56]. The reason for low popularity may be the fact that it is difficult to identify the group of factors which could be universal for all corporate assets. Additionally, as the time goes by, the groups of factors determining risk may change in case of the same assets. Therefore, the selection of factors and analysis of their impact on risk specific for a given enterprise has an arbitral character [57].

After the analysis of the basic assumptions of the models above, it may be concluded that most of them cannot be directly used, without modification, in mining enterprises; furthermore, some of them should even be rejected. These models possess a number of limitations. The primary problem connected with their utilization is the fact that in many countries mining industry does not have many representatives; often the market is ruled by a few enterprises with very high production potential, which causes that this industry has a few representatives on capital market that provide statistical data for the most pricing models based on capital assets. Another problem related to the use of traditional calculation models of return rate is that they usually include systematic risk only and assume that the pricing of return rate is performed from the investor's point of view, who possesses a well-diversified portfolio. Such investor, when calculating the expected return rate, pays attention to market (systematic) risk only. On the other hand, capital provider, who does not possess a diversified investment portfolio, invests a great share of assets in one enterprise; she or he is exposed to both market risk and specific risk of the enterprise. Stock of mining enterprises often, in majority shareholding, belongs to one strategic investor, who is strongly connected with mining industry and does not have a diversified portfolio. The necessity of modification of some models also comes from the issue that they include the elements of specific risk not compliant with the characteristics of mining enterprises or not occurring in the industry. Thus, many categories linked to the return rate calculation in mining enterprises require a new definition or better precision.

6. Modification conception of CAPM for the purpose of expected return rate calculation in mining industry

In this point a modification is suggested regarding the existing method of the expected return rate calculation. An example of the selected mining enterprise—X—is used for that purpose. The expected return rate was calculated for that enterprise and compared with 19 mining enterprises listed on global stock exchanges. The basic subject of research is the mining enterprise from hard coal mining industry, conducting mining activity in Poland, listed on Warsaw Stock Exchange.

For the purpose of research conducted, to enable comparison, the research sample encompassed the largest global mining enterprises in terms of hard coal excavation, comprising the so-called cluster. The parameters characterizing their financial situation are going to form, after an appropriate verification and modifications leading to comparability, benchmarks—reference points for the calculation of expected return rate in Polish mining industry. The selection of enterprises for the research sample was made on the basis of business homogeneity, that is, all the selected enterprises gain a great part of their revenues (over 50%) from hard coal mining activity. Among the benchmark of mining enterprises, the following ones can be mentioned: American corporations (Alpha Natural Resources, Arch Coal, Consol Energy, Peabody Energy Corp., Walter Energy, Inc., Westmoreland Coal Company), Chinese corporations (Inner Mongolia Yitai Coal Company, Ltd., Yanzhou Coal Mining Company, China Coal, Shenhua Group), Australian corporations (BHP Billiton Ltd., Coal of Africa, Coalspur Mines, New Hope Coal, Whitehaven Coal Ltd.), British corporations (Anglo American Coal, Glencore plc), and one Indian corporation (Coal India). They are listed on global stock exchanges such as the New York Stock Exchange (NYSE) in the USA, National Association of Securities Dealers Automated Quotations (NASDAQ)—over the counter, regulated stock market in the USA, Shanghai Stock Exchange (SSE) in China, Hong Kong Stock Exchange (HKSE) in Hong Kong, Australian Securities Exchange (ASX) in Australia, London Stock Exchange (LSE) in Great Britain, and National Stock Exchange of India (NSE) in India.

As it was stated above, the expected return rate depends to a great extent on **specific risk** generated in the particular enterprise. A decisive significance in this area is ascribed to the financial conditions of management process, translating into operational and financial risk of the enterprise. The measures of **operational and financial risk** suggested above were calculated for the selected research sample and placed in **Table 1**. The period of analysis encompassed 5 years. The indicators calculated are listed in a form of average values of the examined parameters in the whole research period, that is, 5 subsequent years.

Some indicators may serve themselves as the measures of specific risk premium of the enterprise. In this work it is assumed that this risk is determined by the whole set of indicators. The first one—degree of operating leverage—identifying the level of operational risk, shows very high fluctuations in the enterprises. DOL fluctuates in the range of over 134 to about 150. Such great volatility with positive or negative numbers appears only in single cases. In general, operating leverage is positive (almost 30% of the examined enterprises are characterized by negative average operating leverage in the examined period, which may mean sales below operational break-even point). In most cases yearly average operational leverage is lower than one, which means that the majority of mining enterprises possesses high vulnerability of operational income to changes in sales revenues. It is specific for hard coal mining industry, where fixed costs dominate. Their great share has an impact on high vulnerability of operating results and on changes in sales amount. The results obtained indicate a high level of operational risk in hard coal mining industry.

The next indicator included in the listing presented in **Table 1** is financial leverage and degree (effect) of financial leverage. Financial leverage is used in most enterprises (apart from one enterprise, all of them use debt capital in financing their activity), and in most cases, we deal with a positive effect of financial leverage. The strongest effects appear in the enterprises with the highest level of debt. In order to determine whether the effect of financial leverage is positive or negative, it was examined whether debt of the enterprises increases or decreases the return on equity. It was observed that the financial leverage is not effectively used, though. It is supported by the fact that, in average, in about half of the enterprises the return on equity is lower than the return on net operating assets. In the majority of the examined enterprises, the

No.	Name	DOL	D/E	DFL	TIE	E/tot capital	D/tot capital	E/fix assets	OC	ROE [%]	RNOA [%}	ROE > RNOA
1	Enterprise X	1.73	0.19	1.03	40.71	0.68	0.13	0.84	+	12.4	10.78	Yes
2	Alpha Natural Resources	−47.72	0.95	1.54	−3.99	0.44	0.40	0.55	+	−18.7	-7.07	No
3	Arch Coal	29.02	1.77	2.62	−0.34	0.36	0.58	0.42	+	−8.5	−0.91	No
4	Consol Energy	0.72	2.11	2.21	5.24	0.30	0.59	0.35	+	14.7	5.32	Yes
5	Peabody Energy Corp.	−0.15	1.66	0.83	1.45	0.34	0.55	0.42	+	3.7	8.13	No
6	Rio Tinto	−153.66	0.94	1.05	5.31	0.45	0.41	0.55	+	8.68	7.61	Yes
7	Walter Energy, Inc.	12.40	3.66	0.92	4.42	0.22	0.71	0.28	+	−7.3	6.80	No
8	Westmorel and Coal Company	13.07	−4.89	−2.42	−0.29	−0.24	1.13	−0.29	—	1.35	2.41	No
9	Inner Mongolia Yitai Coal Company Ltd.	0.98	2.14	1.06	20.83	0.31	0.57	0.41	—	53.1	25.58	Yes
10	Yanzhou Coal Mining Company	134.52	0.97	0.87	24.38	0.42	0.38	0.60	+	16.2	9.93	Yes
11	China Coal	3.81	1.27	1.11	10.91	0.46	0.46	0.69	+	12.0	10.83	Yes
12	Shenhua Group	0.25	0.71	1.04	12.22	0.61	0.43	0.84	—	15.9	20.61	No
13	BHP Billiton Ltd.	−4.54	0.63	1.06	15.68	0.53	0.33	0.68	+	28.6	22.83	Yes
14	Coal of Africa	1.38	0.22	0.99	−116.5	0.79	0.17	0.98	+	−33.0	−11.10	No
15	Coalspur Mines	0.48	0.15	0.96	−19.99	0.83	0.12	1.18	+	−23.0	−27.40	Yes
16	New Hope Coal	−1.83	0.08	1.02	64.27	0.86	0.07	2.96	+	9.83	29.85	No
17	Whitehaven Coal Ltd.	-73.57	0.29	1.13	5.59	0.75	0.22	1.03	+	3.65	2.95	Yes
18	Anglo American Coal	2.54	1.24	1.12	4.77	0.44	0.50	0.57	+	9.45	11.29	No
19	Glencore plc	1.31	5.06	0.50	0.47	0.15	0.62	0.30	—	8.90	3.41	Yes
20	Coal India	−0.17	1.05	1.00	291.99	0.38	0.39	1.94	+	40.3	63.10	No

Source: own work.

Table 1. Basic operational and financial risk measures in the examined enterprises: average values for 5 subsequent years.

cost of servicing debt (TIE) constitutes too big burden for operating income. These problems can be noticed, in particular, in the last 2 years of the analyzed period. Financial risk of the analyzed mining enterprises is growing as the time goes by. An additional risk factor is a low degree of financing fixed assets by equity. It can be seen that in six out of nineteen investigated enterprises equity dominates in their financing structure. These are all Australian enterprises and one Chinese enterprise. Fixed assets in these corporations are to a great extent (60% and more) or even fully covered by equity. In the other cases, it relates to about 2/3 of enterprises, and in their liabilities, structure debt capital dominates; furthermore, a great share of debt capital with interest is clearly outlined, and mostly this capital constitutes over 50% of all liabilities. In this group of enterprises, only about 50% of fixed assets is covered by equity. It means that mining industry, despite a high degree of assets immobilization, finances assets from external sources. Taking the character of fixed assets in mining industry into account, this state of things should be assessed as increasing the risk of financing, which may lead to losing liquidity.

The next examined element is **net operating capital**, also called working capital. Its positive level shows that the part of so-called fixed capital (equity and long-term liabilities), after covering fixed assets, finances current assets. It is a feature of enterprises that use conservative financing strategy, reducing their financing risk. The negative level of working capital means that some part of fixed assets is financed by short-term liabilities. It is a very risky situation from the point of view of losing financial liquidity, and it characterizes the enterprises being in a crisis financial situation or using aggressive strategy, aimed at minimization of the cost of capital. The character of working capital is shown in **Table 1**, marking the sign "+" as positive and the sign "−" as negative net operating capital. The domination of positive or negative working capital is understood as the appearance of the capital with "+" or "−" sign in at least 3 years of the researched period. Negative working capital dominates in the examined period in five mining enterprises. When confronting it with the results obtained in the previous stage, which showed that the share of equity in financing total assets is relatively low in mining industry, it turns out that this industry to a great extent uses financing from long-term debt capital. However, positive working capital indicates that most enterprises in the industry finance fixed assets and some part of current assets by fixed capital, that is, equity and long-term liabilities. Nevertheless, the fact is disturbing that the activity of mining enterprises becomes more risky year by year from the point of view of financial risk. It can be noticed that in the last years of the examined period the number of enterprises with negative working capital is increasing. In the first year of the researched period, only six enterprises note negative working capital, and in the last year, one as many as 11 out of 23. In the last year of analysis, negative working capital in almost half of the examined enterprises shows a very risky configuration, in which highly immobilized fixed assets of mining enterprises, difficult to cash, are financed by short-term liabilities.

Another element taken into account in effectiveness assessment of capital management is **return on equity** and **return on net operating assets**, measured using ROE and RNOA indicators. The analysis of these indicators shows a deteriorating financial situation of mining industry. The return on net operating assets falls down from an average level of about 12% in the first 2 years of analysis to barely 6% in the last researched years. At the beginning of the examined period, only one enterprise from the whole research sample indicates a lack of return on operating assets (Coalspur Mines), and in the last analyzed year, seven enterprises have a negative value of RNOA.

In further analysis ROE is compared with RNOA. A positive evaluation occurs in case of the relation in which the return on equity is higher than the return on net operating assets. The verification of whether ROE > RNOA is made in **Table 1**. The relation of ROE to RNOA, according to the tendency in hard coal mining industry observed before, deteriorates year by year. At the beginning of the examined period only in 1/3 of corporations, the level of ROE is lower than RNOA. In the subsequent years, the number of enterprises characterized by unfavorable relation of these two indicators is increasing, and in the last analyzed year, such situation is noted in almost 80% of the examined enterprises. It means that at the beginning of the examined period in about 65% of the examined enterprises, capital management is effective, that is, the enterprises use a positive effect of financial leverage. However, this situation deteriorated, and, recently, positive assessment of the effects of capital management may obtain only 20% of the investigated corporations.

For the purpose of calculation of the expected return rate in mining enterprises, a model is used in this work. Firstly, a decomposition and qualitative description were made for the most important variables affecting the expected return rate. These are variables representing specific and systematic risk of the enterprise. The next stage of model building is an attempt to parameterize the aforementioned variables.

On the stage of variable selection in the process of model building, cluster analysis is used [58]. Its basis is a selection of research sample among the mining enterprises listed on stock exchanges. On the grounds of the features characterizing the researched areas of enterprise's activity belonging to a particular cluster, the averaged levels of these features were calculated for the whole industry; next, the comparison of averaged values with their values was conducted for the enterprise that the expected return rate was calculated for (in case of this work for the enterprise X). Where necessary, the individual features were brought to comparability with the inclusion of external factors, specific for the enterprises representing different world markets.

Next to specific risk, an important indicator of expected return rate is systematic risk. A classic measure of systematic risk is β coefficient. In case when it cannot be calculated properly, only industry or sector average of β coefficient can be determined and used in CAPM. However, a problem appears that consists in the fact that all the assessed enterprises from the same industry have a common β coefficient, which may be connected with a possible, huge assessment mistake (averaging error).

In some approach it is advised to include, beside systematic risk, the factors of specific risk in addition when calculating β coefficient. Such postulate is proposed by, for example, A. Damodaran, who uses the notion of total risk, marked as "total β," explaining that [28]

$$total\ \beta = industry\ average\ \beta / correlation\ between\ the\ industry\ and\ the\ market \quad (14)$$

Considering the fact that the correlation cannot be higher than 1, total β cannot be lower than a regular one. It allows stating that the inclusion of specific risk always gives a result in a form of increasing the expected return rate [59, 60]. Nevertheless, a problem still remains unsolved concerning the differentiation of the expected return rate for different mining enterprises on the same market. In further analysis an attempt is made to solve this problem through specific risk modeling. It consists in appropriate transformation of β coefficient calculated for similar enterprises listed on stock exchange and adjusting it to the specificity of the investigated corporation. A starting point is a solution suggested in a model by **R. Hamada** [61]. He assumes that risk index is increasing along with the share growth of debt capital in capital structure (providing that other factors remain unchanged). It means that the use of financial leverage in the enterprise increases the risk connected with investing in stock of this enterprise and has a direct translation in the level of β coefficient. According to Hamada's model, β, in the enterprise using financial leverage, may be written as a function of the same indicator in the conditions when debt capital is not used [33]:

$$\beta_{L} = \beta_{U}[1 + (1 - T)(D/E)] \tag{15}$$

where β_L is the β coefficient with financial leverage (levered β), β_U is the β coefficient without financial leverage (unlevered β), and T is the income tax rate.

After transformation this formula takes the following form:

$$B_{U} = \beta_{L}/[1 + (1 - T)(D/E)] \tag{16}$$

The usefulness of both formulas consists in the possibility of making simulation how β is shaping, depending on changes in the level of financial leverage.

In order to analyze systematic risk in the mining enterprise X, as its quotation history on Warsaw Stock Exchange is too short to use statistical methods, cluster analysis was applied. In the enterprises representing global mining industry, β coefficient was calculated. The condition for β determination is compliance of empirical distribution of return rates on stock and market index with normal distribution. In order to examine the empirical distribution of return rates on stock and market indexes, Shapiro–Wilk test and Kolmogorov–Smirnov test with Lilliefors significance correction were used. All the tests univocally indicate that this condition is fulfilled by monthly return rates on stock and market indexes calculated on the basis of 5-year estimation period. Thanks to the selection of enterprises subjected to business homogeneity, grounding on β coefficients calculated for the aforementioned enterprises, industry average β indexes were indicated for the researched market and for the whole research sample in the period of the past 5 years. Industry average β coefficients for the individual markets are included in **Table 2**.

Industry average β for the cluster of hard coal mining in the examined period fluctuates in the range of 1.36–1.54. At the beginning of the analyzed period, the highest value of β is ascribed to the corporations listed in Great Britain (at the beginning this coefficient amounts to 1.92, and after 5 years, it goes down to 1.79). The lowest β coefficient is specific for Chinese enterprises in the whole examined period (average β of enterprises listed on this market equals 0.84). It means that the stock of Chinese mining enterprises is characterized by lower systematic risk than the whole market represented by SSE Composite Index. Beside China, on all the investigated markets, β is higher than 1, which means that the stock of mining enterprises is more risky than the average market portfolio. The stock of British enterprises is the most risky (average β = 1.84) and American ones (average β = 1.66). A bit lower risk is specific for the enterprises listed in Hong Kong (average β = 1.6640) and Australia (average β = 1.43).

In order to include a wider dimension of risk in β coefficient, the correlation between the industry and the market was estimated using Pearson coefficient (the results are placed in **Table 3**). After dividing the industry average β by the correlation coefficient between the industry and the market, total β was obtained, which is shown in **Table 4**.

No.	Name	Average
1	NYSE/NASDAQ (USA)	1.66
2	SSE (China)	0.84
3	HKSE (Hong Kong)	1.40
4	ASX (Australia)	1.43
5	LSE (Great Britain)	1.84
	Total mining industry	1.43

Source: own work.

Table 2. Industry average β coefficient for individual markets with the assumption of 5-year estimation period.

No.	Name	Pearson correlation coefficient
1	NYSE/NASDAQ (USA)	0.632
2	SSE (China)	0.492
3	HKSE (Hong Kong)	0.827
4	ASX (Australia)	0.447
5	LSE (Great Britain)	0.712

Source: own work.

Table 3. Correlation between monthly return rates in the industry and the market.

No.	Name	Total average β
1	NYSE/NASDAQ (USA)	2.62
2	SSE (China)	1.71
3	HKSE (Hong Kong)	1.77
4	ASX (Australia)	2.97
5	LSE (Great Britain)	4.12
6	Total hard coal mining industry	2.70

Source: own work.

Table 4. Total industry average β for individual markets.

The highest correlation between the return rates on stock and the market is noticed in case of the enterprises listed in Hong Kong and Great Britain. The lowest one belongs to stock of Chinese and Australian corporations.

The calculation of total β confirms that the highest risk is specific for stock of British mining enterprises. The lowest risk in the assessed group is characteristic for stock of enterprises listed in Hong Kong and China.

In the area of systematic risk, the subject of analysis was another form of β coefficient. To be exact, β is transformed according to Hamada model, presenting it in an unlevered version for the particular markets (**Table 5**).

Unlevered β coefficient is considered to be a universal systematic risk measure of mining enterprises, and in this form, it is going to be used for modeling of the expected return rate for the enterprises not listed on a capital market or for the enterprises with too short quotation history, based on Polish mining enterprise X.

Specific risk is parameterized based on scoring assessment that includes the operational and financial risk measures presented above. The parameterization of diagnostic features in the enterprise X consists in their comparison with the values of a given feature in the industry, represented by a certain cluster of enterprises and assigning a scoring value to them, corresponding to specific risk class. In **Table 6** there are diagnostic features from scoring model listed, showing at the same time a median of their value for hard coal mining industry (with tolerance level ± 10%), represented by the investigated cluster of enterprises. Median was used in the research as the distribution of the examined diagnostic features is usually asymmetric, which rules out using arithmetic mean.

In the next step, it is suggested to list three specific risk classes, depending on the level of particular diagnostic indicators:

Class 0—low specific risk

Class 1—average specific risk

Class 2—high specific risk

No.	Name	Average unlevered β
1	NYSE/NASDAQ (USA)	0.71
2	SSE (China)	0.48
3	HKSE (Hong Kong)	0.74
4	ASX (Australia)	1.20
5	LSE (Great Britain)	0.99
	Total hard coal mining industry	0.84

Source: own work.

Table 5. Average industry unlevered β for individual markets.

No.	Assessment criterion/diagnostic feature	–10%	Median in global hard coal mining	+10%
1	DOL	1.40	1.55	1.71
2	D/E	0.63	0.70	0.77
3	DFL	0.95	1.05	1.16
4	TIE	4.51	5.01	5.51
5	E/C total	0.40	0.44	0.48
6	D/E total	0.35	0.39	0.43
7	E/fixed assets	0.53	0.59	0.65
8	OC		+	
9	ROE > RNOA		Yes	

Source: own work.

Table 6. Criteria of specific risk assessment with the reference point in a form of median of individual indicators in global hard coal mining.

Each of the accepted diagnostic features is assessed by having assigned the value of 0, 1, or 2, compliant with the specific risk class. The scale was constructed in a descending order, that is, class 0, and assessment 0 is ascribed to the indicators, the values of which is more favorable than the industry average level of a given diagnostic feature, class 1 and assessment 1 (level equal to industry average ± 10%) and class 2 and assessment 2 (unsatisfactory level), meaning a worse level than industry average. The average sum of scores for all diagnostic features is transformed into specific risk premium (SRP). The lower the scoring value, the lower the specific risk and in consequence lower risk premium, which is followed by lower expected return rate.

The first diagnostic feature in a form of the degree of operating leverage informs about the operating risk level. The higher leverage the higher risk. The average industry level of operating leverage in the examined period amounts to 1.55. The enterprises in which the operating leverage is lower than 1.4 (average minus 10%) but higher than 0 are placed in risk class 0. The enterprises with DOL higher than 1.71 (average plus 10%) and a negative one are burdened with high operating risk (risk class 2). If the degree of operating leverage takes the value in the range [1.4; 1.71], the enterprise obtains risk class 1. The next diagnostic features are financial leverage and the degree of financial leverage, informing about the level of financial risk. In average, in the examined cluster of hard coal mining, financial leverage amounts to 0.7 and the degree of financial leverage to 1.05. The increase of these indicators triggers risk growth. When D/E ratio is lower than 0.63 and DFL lower than 0.95, risk class 0 is assigned. If these indicators have the values, D/E from 0.63 to 0.77 and DFL from 0.95 to 1.16, it is risk class 1. Furthermore, above these levels, we deal with risk class 2. Another diagnostic feature is the possibility of debt servicing (TIE). Its higher level is positively evaluated; therefore, the enterprises with the TIE level above 5.51 are in risk class 0, level from 4.51 to 5.51 is risk class 1, and below 4.51 means qualification to risk class 2. Equity to total capital (understood as equity and debt capital with interest) ratio is another criterion of ascribing the enterprise to the particular risk class. The lower the level of this ratio, the higher the risk. The share of equity in total capital below 0.4 is assessed as unsatisfactory and qualifies the particular enterprise to risk class 2. The level from 0.4 to 0.48 is risk class 1 and above 0.48 is class 0. In turn, the next feature—share of debt in total capital—is estimated as unsatisfactory in case of the level below 0.35 (risk class 0). The increase of debt capital in capital structure triggers risk increase. The level of the share of debt capital from 0.35 to 0.43 means risk class 1 and above 0.43 is class 2. The next diagnostic feature relates to the share of equity in financing fixed assets. The lower the scale of financing fixed assets by equity, the higher the risk; thus, the enterprises financing less than 53% of fixed assets by equity are ascribed to risk class 2. Risk class 0 appears in case of financing over 65% of fixed assets by equity. Between these ranges, we deal with, as in previous cases, risk class 1. Another criterion of specific risk assessment is the level of working (operating) capital. For this indicator, a positive level is considered to be satisfactory (risk class 0); indicator equal to zero is risk class 1, and negative working capital means qualifying the enterprise to risk class 2. The last examined criterion is the relation of ROE and RNOA indicators. If ROE > RNOA the corporation is in risk class 0. In adverse situation the relation is unfavorable, which indicates risk class 2. If ROE = RNOA it is risk class 1.

According to the guidelines above, an attempt of parameterization of the assessment criteria of specific risk was made for Polish mining enterprise X (**Table 7**).

Specific risk of enterprise X amounts in the examined period to 0.49 in average. Operational risk increases the level of specific risk (in this area risk class 2 occurs in 3 years of the researched period). In terms of financial risk, the situation of the enterprise is generally more favorable than in the industry, which is confirmed by the domination of financial risk class 0 in the assessment made. However, attention should be paid to the fact that in the last year a rapid deterioration occurred concerning the level of many diagnostic features, which were previously at the more favorable level compared to the industry average. Above all, the ability of the enterprise to service debt is deteriorating, DFL level is increasing, and effectiveness of equity management is decreasing.

No.	Indicator	Year I	Year II	Year III	Year IV	Year V
1	DOL	—	2	0	2	2
2	D/E	0	0	0	0	0
3	DFL	0	1	1	1	2
4	TIE	2	0	0	0	2
5	E/C total	0	0	0	0	0
6	D/E total	0	0	0	0	0
7	E/fixed assets	0	0	0	0	0
8	OC	2	0	0	0	0
9	ROE > RNOA	—	0	0	2	2
	Average score	0.57	0.33	0.11	0.56	0.89

Source: own work.

Table 7. Scoring assessment of diagnostic indicators in mining enterprise X.

No.	Indicator	Year I	Year II	Year III	Year IV	Year V
1.	IRP for hard coal mining [%]	6.00	6.00	6.00	6.00	6.00
2.	$Score_{SR}$	0.57	0.33	0.11	0.56	0.89
3.	SRP [%]	3.43	2.00	0.67	3.33	5.33

Source: own work.

Table 8. Calculation of specific risk premium (SRP) in mining enterprise X.

Having qualified the individual diagnostic features to specific risk classes and performed their scoring assessment, one may start the calculation of specific risk premium. It is assumed that in each enterprise this premium consists of average score of specific risk multiplied by base premium, for which industry premium was adopted, known in practice as *IRP* (*industry risk premium*). The level of IRP was accepted in an arbitral way for the whole examined period based on available research [62–64]. A. Damodaran adopts IRP for hard coal mining industry in the examined period at the level of 5%. In research by B. Francis, I. Hasan, and D. Hunter of American market, the yearly average premium is suggested for mining industry at the level of 7.045%. For the purpose of research conducted in this work, its level was adopted in an arbitral way amounting to 6%.

Specific risk premium in scoring model may be written in the following way:

$$SRP = Score_{SR} \ x \ IRP \quad (17)$$

where SRP is the specific risk premium, $Score_{SR}$ is the arithmetic mean of scores resulting from scoring assessment of specific risk, and IRP is the industry risk premium.

Such construction of specific risk premium means that, when all diagnostic features in the enterprise are in specific risk class 1, specific risk premium is equal to IRP. The enterprises, where the level of diagnostic features is more favorable than the industry average, are characterized by specific risk premium lower than IRP. In turn, the corporations, which have diagnostic features at a more risky level than the industry average, obtain specific risk premium higher than IRP. The listing of specific risk premium (SRP) is included in **Table 8**.

After calculating specific risk premium, one can estimate the expected return rate according to the assumptions of modified CAPM. It may be written as the following formula:

$$C_e = r_{RF} + \beta_U \times (MRP + SRP). \quad (18)$$

7. Conclusions

The calculation of the expected return rate in the Polish mining enterprise X requires the adoption of the following parameters: β coefficient, market risk premium and specific risk premium, and risk-free return rate. The basic parameters necessary for this calculation along with its results are presented in **Table 9**. Risk-free return rate was accepted as the return rate

No.	Indicator	Year I	Year II	Year III	Year IV	Year V
1.	Risk-free rate (10-year treasury bills) [%]	6.17	5.80	5.98	4.94	4.10
2.	Unlevered industry average β	0.82	0.91	0.87	0.85	0.74
3.	MRP [%]	6.08	6.5	7.5	7.3	6.28
4.	SRP [%]	3.43	2.00	0.67	3.33	5.33
5.	Expected return rate [%]	13.97	13.54	13.09	13.98	12.69

Source: own work.

Table 9. Calculation of the cost of equity in Polish mining enterprise X according to the assumptions of modified CAPM.

on 10-year Polish treasury bills. Market risk premium was adopted at the level recommended by A. Damodaran for the whole Polish economy [65]. Furthermore, specific risk premium was accepted at the level according to the calculation made on the basis of scoring model for specific risk parameterization.

The activity of the examined enterprise is burdened with high risk. It is a corporation being in a difficult financial situation. The results obtained may be considered as adequate to risk that is connected with the engagement of capital in this enterprise. The enterprise is in debt and significantly burdened with the cost of debt capital and performs on the edge of financial liquidity. Another big problem is the influence of trade unions on corporate activity and low level of activity diversification. Taking the high risk into account, capital providers expect the return rate at the level of about 13%.

The presented calculation model of the expected return rate does not lack limitations. In comparison with the traditional CAPM, it is better adjusted to the specificity of the activity of mining enterprises. Its limitation may be some kind of subjectivism, especially in relation with the specific risk factors that are difficult to measure. However, the solution proposed eliminates the basic defects of the existing calculation models of the expected return rate, which made it impossible to obtain a realistic pricing in mining enterprises. The models suggested constitute a significant help for the managers of mining enterprises.

Author details

Aneta Michalak

Address all correspondence to: aneta.michalak@polsl.pl

Silesian University of Technology, Faculty of Organization and Management, Gliwice, Poland

References

[1] Dobija M, Dobija D. O naturze kapitału. Theoretical Accounting Research Bulletins. 2003;**17**:7

[2] Bohm-Bawerk E. Capital and Interest, the Controversy over the Concept of Capital. South Holland, Illinois: Libertarian Press; 1959. p. 152

[3] de Soto H. Tajemnica kapitału: dlaczego kapitalizm triumfuje na Zachodzie a zawodzi gdzie indziej. Chicago, Warsaw: Fijor Publishing; 2002. p. 286

[4] Bernstein PL. Capital Ideas Evolving. Hoboken, New Jersey: J. Wiley & Sons, Inc.; 2007. p. 304. DOI: 10.2469/faj.v64.n4.8

[5] Kortum S, Lerner J. Assessing the contribution of venture capital to innovation. Rand Journal of Economics. 2000;**31**(4):674-678. DOI: 10.2307/2696354

[6] Janasz K. Kapitał w finansowaniu działalności innowacyjnej przedsiębiorstw w Polsce. Źródła i modele. Warsaw: Difin Publishing S.A; 2010. p. 280

[7] Borowiecki R. Zarządzanie kapitałem trwałym przedsiębiorstwa w gospodarce rynkowej. Szczecin: University of Szczecin Publishing; 1993. p. 119

[8] Dobija M. Ewolucja koncepcji rachunkowości, kapitału i pieniądza. In: Sojak S, editor. Historia, współczesność i perspektywy rachunkowości w. Polsce, Toruń: Nicolaus Copernicus University in Toruń Publishing; 2003. p. 920

[9] Duliniec A. Struktura i koszt kapitału w przedsiębiorstwie. Warsaw: PWN Publishing; 2001. p. 173

[10] Ross SA, Westerfield RW, Jordan BD. Fundamentals of Corporate Finance. Chicago: Mcgraw-Hill/Irwin; 1996. p. 753

[11] Ogier T, Rugman J, Spicer L. The Real Cost of Capital: A Business Field Guide to better Financial Decisions. Pearson Education: Glasgow; 2004. p. 320

[12] Byrka-Kita K. Metody szacowania kosztu kapitału własnego. Teoria a praktyka. University of Szczecin Theses and Studies. 2008;**689**:26

[13] Haugen R, Senbet L. The insignificance of bankruptcy costs to the theory of optimal capital structure. Journal of Finance. 1978;**33**(2):383-393. DOI: 10.1111/j.1540-6261.1978.tb04855.x

[14] Michalak A. Conceptualization of the cost of capital in the theory of economy and contemporary economic sciences. In: SGEM Conference on Political Sciences, Law, Finance, Economics and Tourism. SGEM 2014, Albena, Bulgaria, September 1-9, 2014. Conference proceedings. Vol. 2. Sofia: STEF92 Technology; 2014, pp. 115-120. DOI: 10.5593/sgemsocial2014/b22/s6.015

[15] Gospodarowicz A. Ryzyko operacyjne i jego ocena w regulacjach Nowej Umowy Kapitałowej. In: Jajuga K, editor. Wyzwania współczesnych finansów. Wrocław: Wrocław University of Economics Publishing; 2009. p. 28

[16] Thlon M. Zarządzanie ryzykiem operacyjnym przedsiębiorstwa: metoda szacowania ryzyka delta-EVT. Kraków: University of Economics Publishing; 2012. p. 175

[17] Jaruga A, Nowak W, Szychta A. Rachunkowość zarządcza. Koncepcje i zastosowania. Łódź: Absolwent Publishing; 2000. p. 926

[18] Wirth H, Wanielista K, Burta J, Kicki J. Strategiczna I Ekonomiczna Ocena przemysłowych projektów Inwestycyjnych. Kraków: IGSMiE PAN Publishing; 2000. p. 160

[19] Jonek-Kowalska I, Turek M, editors. Zarządzanie ryzykiem operacyjnym w przedsiębiorstwie górniczym. Warsaw: PWN Publishing; 2011. p. 418

[20] Michalak A. Modele kosztu kapitału i ich implementacje w zarządzaniu przedsiębiorstwem górniczym. Warsaw: Difin Publishing; 2015. p. 344

[21] Wirth H. Wieloczynnikowa wycena złóż I ich zasobów na przykładzie przemysłu metali nieżelaznych. Kraków: IGSMiE PAN Publishing; 2011. p. 208

[22] Stretcher R, Johnson S. Capital structure: Professional management guidance. Managerial Finance. 2001;**8**(37):788-804. DOI: 10.1108/03074351111146229

[23] Blume ME, Friend I, Westerfield R. Impediments to Capital Formation. Philadelphia: University of Pensylvania, Rodney White Center for Financial Research, The Warton School; 1980. p. 63

[24] Gabrusewicz W. Analiza finansowa przedsiębiorstwa: teoria i zastosowanie. Warsaw: PWE Publishing; 2014. p. 376

[25] Sierpińska M, Jachna T. Ocena przedsiębiorstwa według standardów światowych. Warsaw: PWN Publishing; 2006. p. 255

[26] Kurek B. Estymacja premii za ryzyko na podstawie ROE. In: Standardy edukacyjne rachunkowości: praktyka i stan badań. Vol. 1079. Wrocław: Scientific Theses of Wrocław University of Economics; 2005. p. 230

[27] Michalak A. Koszt Kjako Przesłanka Doboru źródeł Finansowania Działalności Operacyjnej. In: Turek M, editor. Modele Finansowania działalności Operacyjnej przedsiębiorstw górniczych. Katowice: Central Mining Institute; 2011. p. 436

[28] Bernstein P, Damodaran A. Zarządzanie inwestycjami. Warsaw: LIBER Publishing; 1999. p. 451

[29] Melich M Wycena wartości firmy. In: Szablewski A, Tuzimek R, editors. Wycena I zarządzanie wartością Firmy. Warsaw: PolteP1t Publishing; 2004. pp. 137-291

[30] Mayo HB. Wstęp do inwestowania. Warsaw: LIBER Publishing; 1997. p. 909

[31] Rakow KC. The effect of management earning forecast characteristics on cost of equity capital. Advances in Accounting, Incorporating Advances in International Accounting. 2010;**26**:37-46. DOI: 10.1016/j.adiac.2010.02.007

[32] Alexander I. Cost of Capital. The Application of Financial Models to State Aid. Oxford: OXERA; 1995. p. 102

[33] Damodaran A. Estimating Risk Parameters [Internet]. 2012. New York: Stern School of Business; Available from: http://archive.nyu.edu/bitstream/2451/26906/3/wpa99019.pdf.txt [Accessed: October 21, 2016]

[34] Sharpe W. Capital asset prices: A theory of market equilibrium. Journal of Finance. 1964;**19**(3):425-442. DOI: 10.1111/j.1540-6261.1964.tb02865.x

[35] Linther J. The valuation of risk assets and the selection of risky investments in stock portfolios and capital budgets. Review of Economics and Statistics. 1965;**47**(1):13-37. DOI: 10.2307/1924119

[36] Kapil S. Financial Management. Dorling Kinderlsey, India, licensees of Pearson Education in South Asia: New Delhi; 2011. p. 752

[37] Zarzecki D. Metody wyceny przedsiębiorstw. Fundacja Rozwoju Rachunkowości: Warsaw; 1999. p. 205

[38] Pratt SP, Grabowski RJ, editors. Cost of Capital: Applications and Examples. Hoboken, New Jersey: Wiley; 2014. p. 1344. DOI: 10.1002/9781118846780

[39] Brigham EF, Gapenski LC. Zarządzanie finansami, t. 1. Warsaw: PWE; 2000. p. 892

[40] Buckley A. Inwestycje zagraniczne. Składniki, wartości i ocena. Warsaw: PWN; 2002. p. 268

[41] Berk J, DeMarzo P. Corporate Finance. 2nd ed. Boston: Pearson; 2011. p. 1001

[42] Włoszczowski B. Krytyczne uwagi na temat metod oceny projektów inwestycyjnych. In: Dura A, editor. Budżetowanie działalności jednostek gospodarczych: Teoria i praktyka. Kraków: AGH; 2012. pp. 188-194

[43] Kidwell DS, Moles P, Parrino R. Fundamentals of Corporate Finance: European Edition. John Wiley & Sons Ltd.: Chichester; 2011. p. 706

[44] Hawawini G, Viallet C. Finance for Executives: Managing for Value Creation. 4th ed. Mason: South-Western Cengage Learning; 2011. p. 672

[45] Damodaran A. Investment Valuation. Tools and Techniques for Determining the Value of any Asset. Hoboken, New Jersey: John Wiley & Sons Inc., Wiley Frontiers in Finance; 2012. p. 974

[46] Williams JB. The Theory of Investment Value. Cambridge, Massachusetts: Harvard University Press; 1938. p. 637

[47] Gordon MJ, Shapiro E. Capital equipment analysis: The required rate of profit. Management Science. 1956;**3**(1):102-110. DOI: 10.1287/mnsc.3.1.102

[48] Black AJ. Macroeconomic risk and the Fama-French three-factor model. Managerial Finance. 2006;**32**(6):505-517. DOI: 10.1108/03074350610666238

[49] Kothari SP, Shanken J, Sloan RG. Another look at the cross-section of expected stock returns. The Journal of Finance. 1995;**50**(1):185-224. DOI: 10.1111/j.1540-6261.1995.tb05171.x

[50] Black F. Beta and return. The Journal of Portfolio Management. 1993;**20**(1):8-18. DOI: 10.3905/jpm.1993.409462

[51] Chan LK, Lakonishok J. Are the Reports of β's Death Premature? Working Paper. Illinois: University of Illinois; 1992. pp. 51-61

[52] Chiu KC, Xu L. Arbitrage pricing theory-based Gaussian temporal factor analysis for adaptive portfolio management. Decision Support Systems. 2004;**37**(4):485-500. DOI: 10.1016/s0167-9236(03)00082-4

[53] Magni CA. Project valuation and investment decisions: CAPM versus arbitrage. Applied Financial Economics Letters. 2007;**3**(2):137-140. DOI: 10.1080/17446540500426821

[54] Reisman H. Some comments on the APT. Quantitative Finance. 2002;**2**(5):378-386. DOI: 10.1088/1469-7688/2/5/307

[55] Balbás A, Balbás B, Balbás R. CAPM and APT-like models with risk measures. Journal of Banking & Finance. 2010;**34**(6):1166-1174. DOI: 10.1016/j.jbankfin.2009.11.013

[56] Zarzecki D. Techniki szacowania kosztu kapitału własnego. Ekonomika i Organizacja Przedsiębiorstwa. 2000;**8**:18-20

[57] Dimson E, Mussavian M. Three centuries of asset pricing [internet]. SSRN Electronic Journal. Journal of Banking and Finance. 1999;**23**(12):1745-1769. LBS Institute of Finance and Accounting Working Paper No. IFA 385. Posted 11th January 2000. DOI: 10.2139/ssrn.203108.www.papers.ssm.com [Accessed: June 9, 2014]

[58] Ingram M, Margetis S. A practical method to estimate the cost of equity for a firm using cluster analysis. Managerial Finance. 2010;**2**(36):160-167. DOI: 10.1108/03074351011014569

[59] Butler PJ. Total Beta, who's right, who's wrong [internet]. Valuation Strategies. WG&L Journals. February 2011. Available from: www.bvmarketdata.com/default-textonly.asp?f=bpmarticles [Accessed: March 20, 2015]

[60] Byrka-Kita K. Błędy w szacowaniu kosztu kapitału: Analiza wybranych problemów. University of Szczecin Research Bulletins. 2007;**455**(1):267

[61] Hamada RS. The effect of the firm's capital structure on the systematic risk of common stock. Journal of Finance. 1972;**2**(27):435-452. DOI: 0.2307/2978486

[62] Cappiello L, Duca ML, Maddaloni A. Country and industry equity risk premia in the euro area. European Central Bank, Working Paper Series. 2008;**913**:23

[63] Francis B, Hasan I, Hunter D. Does hedging tell the full story? Reconciling differences in US aggregate and industry: Level exchange rate risk premia. SSRN Electronic Journal. 2008. DOI: 10.2139/ssrn.1138148

[64] Damodaran A. Equity risk premiums [internet]. 2010-2014. New York: Stern School of Business. Available from: http://www1.worldbank.org/finance/assets/images/Equity_Risk_Premiums.pdf [Accessed: May 20, 2015]

[65] Damodaran A. Bond and spread ratings [internet]. 2010-2014. New York: Stern School of Business. Available from: http://pages.stern.nyu.edu/~adamodar/New_Home_Page/datafile/ctryprem.html [Accessed: March 19, 2015]